Meaning and Necessity

A Study in Semantics and Modal Logic

By

RUDOLF CARNAP

THE UNIVERSITY OF CHICAGO PRESS

CHICAGO AND LONDON

International Standard Book Number: 0-226-09346-8

THE UNIVERISTY OF CHICAGO PRESS, CHICAGO 60637

The University of Chicago Press, Ltd., London

 Published 1947. Enlarged Edition, 1956. First Phoenix Edition 1958. Sixth Impression 1970. Printed in the United States of America.

PREFACE TO THE FIRST EDITION

The main purpose of this book is the development of a new method for the semantical analysis of meaning, that is, a new method for analyzing and describing the meanings of linguistic expressions. This method, called *the method of extension and intension*, is developed by modifying and extending certain customary concepts, especially those of class and property. The method will be contrasted with various other semantical methods used in traditional philosophy or by contemporary authors. These other methods have one characteristic in common: They all regard an expression in a language as a name of a concrete or abstract entity. In contradistinction, the method here proposed takes an expression, not as naming anything, but as possessing an intension and an extension.

This book may be regarded as a third volume of the series which I have called "Studies in Semantics", two volumes of which were published earlier. However, the present book does not presuppose the knowledge of its predecessors but is independent. The semantical terms used in the present volume are fully explained in the text. The present method for defining the L-terms (for example, 'L-true', meaning 'logically true', 'analytic') differs from the methods discussed in the earlier *Introduction to Semantics*. I now think that the method used in this volume is more satisfactory for languages of a relatively simple structure.

After meaning analysis, the second main topic discussed in this book is *modal logic*, that is, the theory of modalities, such as necessity, contingency, possibility, impossibility, etc. Various systems of modal logic have been proposed by various authors. It seems to me, however, that it is not possible to construct a satisfactory system before the meanings of the modalities are sufficiently clarified. I further believe that this clarification can best be achieved by correlating each of the modal concepts with a corresponding semantical concept (for example, necessity with L-truth). It will be seen that this method also leads to a clarification and elimination of certain puzzles which logicians have encountered in connection with modalities. In the Preface to the second volume of "Studies in Semantics," I announced my intention to publish, as the next volume, a book on modal logic containing, among other things, syntactical and semantical systems which combine modalities with quantification. The present book, however, is not as yet the complete fulfilment of that promise: it contains only analyses and discussions of modalities, preliminary to the construction of modal systems. The systems themselves are not given here. In an article published elsewhere (see Bibliography), I have stated a calculus and a semantical system combining modalities with quantification, and have summarized some of the results concerning these systems. A more comprehensive exhibition of results already found and those yet to be found must be left for another time.

The investigations of modal logic which led to the methods developed in this book were made in 1942, and the first version of this book was written in 1943, during a leave of absence granted by the University of Chicago and financed by the Rockefeller Foundation. To each of these institutions I wish to express my gratitude for their help. Professors Alonzo Church and W. V. Quine read the first version and discussed it with me in an extensive correspondence. I am very grateful to both for the stimulation and clarification derived from this discussion, and to Quine also for a statement of his view and, in particular, of his reaction to my method of modal logic. This statement is quoted in full and discussed in detail in the penultimate section of this book. I am also indebted to Professors Carl G. Hempel and J. C. C. McKinsey for some helpful comments. To Miss Gertrude Jaeger I am grateful for expert help in the preparation of the manuscript.

R. C.

CHICAGO

PREFACE TO THE SECOND EDITION

The main body of this book is unchanged. But a Supplement is added containing five previously published articles. They grew out of discussions about problems dealt with in this book. They sometimes give a more detailed or clearer formulation of my position, sometimes they represent a modification of my earlier views, often stimulated by discussions and objections by other authors.

The contents of Articles A–E in the Supplement are related to certain sections of the book, as follows. Article B outlines a new method in connection with the definition of L-truth in § 2 and the conception of state-descriptions as descriptions of possible states. The problem of the nature and admissibility of propositions and other entities discussed in §§ 6 and 10 is dealt with in greater detail in Article A. Article C indicates a change in the explication of belief-sentences given in §§ 13–15. Article D defends the semantical concept of intension against extensionalist objections, like those of Quine discussed in § 44, by showing the scientific legitimacy of the corresponding pragmatical concept of linguistic meaning. Article E adds to this some brief remarks on pragmatical concepts.

Many references to more recent publications have been added to the Bibliography.

I wish to thank the editors of the *Revue Internationale de Philosophie*, the *Philosophical Studies*, and the publisher (Basil Blackwell, Oxford) of the book *Philosophy and Analysis* for their kind permission to reprint the articles.

RUDOLF CARNAP

UNIVERSITY OF CALIFORNIA AT LOS ANGELES

CONTENTS

CHAPTER I

THE METHOD OF EXTENSION AND INTENSION

A method of semantical meaning analysis is developed in this chapter. It is applied to those expressions of a semantical system S which we call *designators;* they include (declarative) sentences, individual expressions (i.e., individual constants or individual descriptions) and predicators (i.e., predicate constants or compound predicate expressions, including abstraction expressions). We start with the semantical concepts of *truth* and *L-truth* (logical truth) of sentences (§§ 1, 2). It is seen from the definition of L-truth that it holds for a sentence if its truth follows from the semantical rules alone without reference to (extra-linguistic) facts (§ 2). Two sentences are called (materially) *equivalent* if both are true or both are not true. The use of this concept of equivalence is then extended to designators other than sentences. Two individual expressions are equivalent if they stand for the same individual. Two predicators (of degree one) are equivalent if they hold for the same individuals. *L-equivalence* (logical equivalence) is defined both for sentences and for other designators in such a manner that it holds for two designators if and only if their equivalence follows from the semantical rules alone. The concepts of equivalence and L-equivalence in their extended use are fundamental for our method (§ 3).

If two designators are equivalent, we say also that they have the same *extension*. If they are, moreover, L-equivalent, we say that they have also the same *intension* (§ 5). Then we look around for entities which might be taken as extensions or as intensions for the various kinds of designators. We find that the following choices are in accord with the two identity conditions just stated. We take as the extension of a predicator the class of those individuals to which it applies and, as its intension, the property which it expresses; this is in accord with customary conceptions (§ 4). As the extension of a sentence we take its truth-value (truth or falsity); as its intension, the proposition expressed by it (§ 6). Finally, the extension of an individual expression is the individual to which it refers; its intension is a concept of a new kind expressed by it, which we call an individual concept (§§ 7–9). These conceptions of extensions and intensions are justified by their fruitfulness; further definitions and theorems apply equally to extensions of all types or to intensions of all types.

A sentence is said to be *extensional* with respect to a designator occurring in it if the extension of the sentence is a function of the extension of the designator, that is to say, if the replacement of the designator by an equivalent one transforms the whole sentence into an equivalent one. A sentence is said to be *intensional* with respect to a designator occurring in it if it is not extensional and if its intension is a function of the intension of the designator, that is to say, if the replacement of this designator by an L-equivalent one transforms the whole sentence into an L-equivalent one. A modal sentence (for example, 'it is necessary that . . .') is intensional with respect to its subsentence (§ 11). A psychological sentence like 'John believes that it is raining now' is neither extensional nor intensional with respect to its subsentence (§ 13). The problem of the semantical analysis of these *belief-sentences* is solved with the help of the concept of *intensional structure* (§§ 14, 15).

§ 1. Preliminary Explanations

This section contains explanations of a symbolic language system S_1, which will later serve as an *object language* for the illustrative application of the semantical methods to be discussed in this book. Further, some semantical concepts are explained for later use; they belong to the semantical *metalanguage* M, which is a part of English. Among them are the concepts of *truth, falsity*, and (material) *equivalence*, applied to sentences. The term '*designator*' is introduced for all those expressions to which a semantical meaning analysis is applied, the term will be used here especially for sentences, predicators (i.e., predicate expressions), and individual expressions.

The chief task of this book will be to find a suitable method for the semantical analysis of meaning, that is, to find concepts suitable as tools for this analysis. The concepts of the intension and the extension of an expression in language will be proposed for this purpose. They are analogous to the customary concepts of property and class but will be applied in a more general way to various types of expressions, including sentences and individual expressions. The two concepts will be explained and discussed in chapters i and ii.

The customary concept of name-relation and the distinction sometimes made since Frege between the entity named by an expression and the sense of the expression will be discussed in detail in chapter iii. The pair of concepts, extension-intension, is in some respects similar to the pair of Frege's concepts; but it will be shown that the latter pair has serious disadvantages which the former avoids. The chief disadvantage of the method applying the latter pair is that, in order to speak about, say, a property and the corresponding class, two different expressions are used. The method of extension and intension needs only one expression to speak about both the property and the class and, generally, one expression only to speak about an intension and the corresponding extension.

In chapter iv, a metalanguage will be constructed which is neutral with regard to extension and intension, in the sense that it speaks not about a property and the corresponding class as two entities but, instead, about one entity only; and analogously, in general, for any pair of an intension and the corresponding extension. The possibility of this neutral language shows that our distinction between extension and intension does not presuppose a duplication of entities.

In chapter v, some questions concerning modal logic are discussed on the basis of the method of extension and intension.

My interest was first directed toward the problems here discussed when I was working on systems of modal logic and found it necessary to clarify the concepts which will be discussed here under the terms of 'extension' and 'intension' and related concepts which have to do with what is usually called the values of a variable. Further stimulation came from some recent publications by Quine[1] and Church,[2] whose discussions are valuable contributions to a clarification of the concepts of naming and meaning.

Before we start the discussion of the problems indicated, some explanations will be given in this section concerning the object languages and the metalanguage to be used. We shall take as object languages mostly symbolic languages, chiefly three semantical language systems, S_1, S_2, and S_3, and occasionally also the English word language. For the sake of brevity, not all the rules of these symbolic systems will be given, but only those of their features will be described which are relevant to our discussion. S_1 will now be described; S_2 is an extension of it that will be explained later (§ 41); S_3 will be described in § 18.

The *system S_1* contains the customary connectives of negation '$\sim$' ('not'), disjunction '$\vee$' ('or'), conjunction '$\bullet$' ('and'), conditional (or material implication) '$\supset$' ('if . . . then . . .'), and biconditional (or material equivalence) '$\equiv$' ('if and only if'). The only variables occurring are individual variables 'x', 'y', 'z', etc. For these variables the customary universal and existential quantifiers are used: '$(x)(\ldots x \ldots)$' ('for every x, . . x . .') and '$(\exists x)(\ldots x \ldots)$' ('there is an x such that . . x . .'). All sentences in S_1 and the other systems are closed (that is, they do not contain free variables). In addition to the two quantifiers, two other kinds of operators occur: the iota-operator for individual descriptions ('$(\iota x)(\ldots x \ldots)$', 'the one individual x such that . . x . .') and the lambda-operator for abstraction expressions ('$(\lambda x)(\ldots x \ldots)$', 'the property (or class) of those x which are such that . . x . .'). If a sentence consists of an abstraction expression followed by an individual constant, it says that the individual has the property in question. Therefore, '$(\lambda x)(\ldots x \ldots)$a' means the same as ' . . a . .', that is, the sentence formed from '. . x . .' by substituting 'a' for 'x'. The rules of our system will permit the transformation of '$(\lambda x)(\ldots x \ldots)$a' into '. . a . .' and vice versa; these transformations are called conversions.

S_1 contains descriptive constants (that is, nonlogical constants) of indi-

[1] [Notes] (see Bibliography at the end of this book). Quine's views concerning the name-relation (designation) will be discussed in chap. iii; and the conclusions which he draws from them for the problem of quantification in modal sentences will be discussed in chap. v.

[2] [Review C.] and [Review Q.]. Church's conceptions will be discussed in chap. iii, in connection with those of Frege. Church's contributions are more important than is indicated by the form of their publication as reviews. It is to be hoped that he will soon find the opportunity for presenting his conception in a more comprehensive and systematic form.

vidual and predicate types. The number of predicates in S_1 is supposed to be finite, that of individual constants may be infinite. For some of these constants, which we shall use in examples, we state here their meanings by semantical rules which translate them into English.

1-1. *Rules of designation for individual constants*
's' is a symbolic translation of 'Walter Scott',
'w'—'(the book) Waverley'.

1-2. *Rules of designation for predicates*
'Hx'—'x is human (a human being)',
'RAx'—'x is a rational animal',
'Fx'—'x is (naturally) featherless',
'Bx'—'x is a biped',
'Axy'—'x is an author of y'.

The English words here used are supposed to be understood in such a way that 'human being' and 'rational animal' mean the same. Further, we shall use 'a', 'b', 'c', as individual constants, and 'P', 'Q', as predicator constants (of level one and degree one); the interpretation of these signs will be specified in each case, or left unspecified if not relevant for the discussion.

In order to speak *about* any *object language*—here the symbolic language systems S_1, etc.—we need a *metalanguage*. We shall use as our metalanguage M a suitable part of the English language which contains translations of the sentences and other expressions of our object languages (for example, the translations stated in 1-1 and 1-2), names (descriptions) of those expressions, and special semantical terms. For the sake of simplicity, we shall usually construct a name of an expression in the customary way by including it in single quotation marks. In order to speak about expressions in a general way, we often use '$\mathfrak{A}_i$', '$\mathfrak{A}_j$', etc., for expressions of any kind and '$\mathfrak{S}_i$', '$\mathfrak{S}_j$', etc., for sentences, sometimes also blanks like '. . .', '- -', etc., and blanks with a variable, e.g., '. . x . .', for an expression in which that variable occurs freely. If a German letter occurs in an expression together with symbols of the object language, then the latter ones are used autonymously, i.e., as names for themselves.[3] Thus, we may write in M, for instance, '$\mathfrak{A}_i \equiv \mathfrak{A}_j$'; this is meant as referring to that expression of the object language which consists of the expression $\mathfrak{A}_i$ (whatever this may be, e.g., 'Hs') followed by the sign '$\equiv$', followed by the expression $\mathfrak{A}_j$. (In symbolic formulas both in the object languages and in M, parentheses will often be omitted under the customary conditions.) The term

[3] See [Syntax], § 42.

'***sentence***' will be used in the sense of 'declarative sentence'. The term 'sentential matrix' or, for short, '***matrix***' will be used for expressions which are either sentences or formed from sentences by replacing individual constants with variables. (If a matrix contains any number of free occurrences of *n* different variables, it is said to be of degree *n;* for example, 'A*xy* V P*x*' is of degree two; the sentences are the matrices of degree zero). A sentence consisting of a predicate of degree *n* followed by *n* individual constants is called an ***atomic sentence*** (e.g., 'Pa', 'Abc').

A complete construction of the semantical system S_1, which cannot be given here, would consist in laying down the following kinds of rules: (1) rules of formation, determining the admitted forms of sentences; (2) rules of designation for the descriptive constants (e.g. 1-1 and 1-2); (3) rules of truth, which we shall explain now; (4) rules of ranges, to be explained in the next section. Of the ***rules of truth*** we shall give here only three examples, for atomic sentences (1-3), for 'V' (1-5), and for '≡' (1-6).

1-3. *Rule of truth* for the simplest atomic sentences. An atomic sentence in S_1 consisting of a predicate followed by an individual constant is true if and only if the individual to which the individual constant refers possesses the property to which the predicate refers.

This rule presupposes the rules of designation. It yields, together with rules 1-1 and 1-2, the following result as an example:

1-4. The sentence 'Bs' is true if and only if Scott is a biped.

1-5. *Rule of truth* for 'V'. A sentence $\mathfrak{S}_i \vee \mathfrak{S}_j$ is true in S_1 if and only if at least one of the two components is true.

1-6. *Rule of truth* for '≡'. A sentence $\mathfrak{S}_i \equiv \mathfrak{S}_j$ is true if and only if either both components are true or both are not true.

There are some further rules of truth for the other connectives, corresponding to their truth-tables, and for the quantifiers; another example of a rule of truth will be given in 3-3. The rules of truth together constitute a recursive *definition for* '*true in* S_1', because they determine, in combination with the rules of designation, for every sentence in S_1 a sufficient and necessary condition of its truth (as is given for 'Bs' in 1-4). Thereby they give an *interpretation* for every sentence. Thus, for example, we learn from the rules that the sentence 'Bs' says that (in other words, expresses the proposition that) Scott is a biped. For the purposes of our discussion it is not necessary to give the whole definition of truth.[4] It will suffice to pre-

[4] The first definition of the semantical concept of truth was given by Tarski [Wahrheitsbegriff]; I have given a slightly different form in [I], § 7. For nontechnical discussions of the nature of the semantical concept of truth see Tarski [Truth] and my [Remarks].

suppose that the term 'true' is defined in such a manner that it has its customary meaning as applied to sentences. More specifically, we presuppose that a statement in M saying that a certain sentence in S_1 is true means the same as the translation of this sentence;[5] for example, 'the sentence 'Hs' is true in S_1' means the same as 'Walter Scott is human'. On the basis of 'true', some further semantical terms are defined as follows, with respect to any semantical system *S*, e.g., S_1, etc.

* **1-7.** *Definition.* $\mathfrak{S}_i$ is ***false*** (in *S*) $=_{Df} \sim \mathfrak{S}_i$ is true (in *S*).

Thus 'false' has here its ordinary meaning.

1-8. *Definition.* $\mathfrak{S}_i$ is ***equivalent*** to $\mathfrak{S}_j$ (in *S*) $=_{Df} \mathfrak{S}_i \equiv \mathfrak{S}_j$ is true (in *S*).

This definition, together with the rule of truth for '≡' (1-6), yields this result:

1-9. Two sentences are equivalent if and only if both have the same truth-value, that is to say, both are true or both are false.

It is to be noticed that the term 'equivalent' is here defined in such a manner that it means merely agreement with respect to truth-value (truth or falsity), a relation which is sometimes called 'material equivalence'. The term is here not used, as in ordinary language, in the sense of agreement in meaning, sometimes called 'logical equivalence'; for the latter concept we shall later introduce the term 'L-equivalent' (2-3c).

I propose to use the term ***'designator'*** for all those expressions to which a semantical analysis of meaning is applied, the class of designators thus being narrower or wider according to the method of analysis used. [The word 'meaning' is here always understood in the sense of 'designative meaning', sometimes also called 'cognitive', 'theoretical', 'referential', or 'informative', as distinguished from other meaning components, e.g., emotive or motivative meaning. Thus here we have to do only with declarative sentences and their parts.] Our method takes as designators at least sentences, ***predicators***[6] (i.e., predicate expressions, in a wide sense,

[5] For detailed discussions of this characteristic of the semantical concept of truth, see Tarski [Truth] and my [Remarks], § 3.

[6] Some terms with the ending '-tor' for kinds of expressions are customary, e.g., 'functor', 'operator'. The terms 'predicator' and 'designator' are formed in analogy to them. A still wider use of the same ending might be taken into consideration with the aim of making the terminology in the metalanguage somewhat more uniform. For this book, only the two terms mentioned are adopted; but the following terms would seem to me quite suitable, too: 'descriptor' (for the customary 'description'), 'abstractor' (for 'abstraction expression'), 'connector' (for 'connective'). Other terms might seem more questionable, but perhaps still worth consideration, e.g., 'individuator' (for 'individual expression'), 'propositor' or 'stator' (for '(declarative) sentence'), 'conceptor' (for 'concept expression,' i.e., 'designator other than sentence'). Morris, [Signs], uses a number of terms with '-tor' (or '-or'), among them some of those mentioned here, for kinds of expressions or, more generally, of signs including non-linguistic signs.

including class expressions), functors (i.e., expressions for functions in the narrower sense, excluding propositional functions), and individual expressions; other types may be included if desired (e.g., connectives, both extensional and modal ones). The term 'designator' is not meant to imply that these expressions are names of some entities (the name-relation will be discussed in § 24), but merely that they have, so to speak, an independent meaning, at least independent to some degree. Only (declarative) sentences have a (designative) meaning in the strictest sense, a meaning of the highest degree of independence. All other expressions derive what meaning they have from the way in which they contribute to the meaning of the sentences in which they occur. One might perhaps distinguish—in a vague way—different degrees of independence of this derivative meaning. Thus, for instance, I should attribute a very low degree to '(', somewhat more independence to '∨', still more to '+' (in an arithmetical language), still more to 'H' ('human') and 's' ('Scott'); I should not know which of the last two to rank higher. This order of rank is, of course, highly subjective. And where to make the cut between expressions with no or little independence of meaning ('syncategorematic' in traditional terminology) and those with a high degree of independence, to be taken as designators, seems more or less a matter of convention. If a metalanguage is decided upon, then it seems convenient to take as designators at least the expressions of all those types, but not necessarily only those, for which there are variables in the metalanguage (compare [I], § 12, and references to Quine, below, at the beginning of § 10).

§ 2. L-Concepts

By the *explication* of a familiar but vague concept we mean its replacement by a new exact concept; the former is called explicandum, the latter explicatum. The concept of *L-truth* is here defined as an explicatum for what philosophers call logical or necessary or analytic truth. The definition leads to the result that a sentence in a semantical system is L-true if and only if the semantical rules of the system suffice for establishing its truth. The concepts of L-falsity, L-implication, and *L-equivalence* are defined as explicata for logical falsity, logical implication or entailment, and mutual logical implication, respectively. A sentence is called *L-determinate* if it is either L-true or L-false; otherwise it is called L-indeterminate or *factual*. The latter concept is an explicatum for what Kant called synthetic judgments. A sentence is called *F-true* if it is true but not L-true; F-truth is an explicatum for what is known as factual or synthetic or contingent truth. The concepts of F-falsity, F-implication, and F-equivalence are defined analogously.

The task of making more exact a vague or not quite exact concept used in everyday life or in an earlier stage of scientific or logical development, or rather of replacing it by a newly constructed, more exact concept, belongs among the most important tasks of logical analysis and logical construction. We call this the task of explicating, or of giving an ***explication*** for, the earlier concept; this earlier concept, or sometimes the term used for it, is called the ***explicandum;*** and the new concept, or its term, is called an ***explicatum*** of the old one.[7] Thus, for instance, Frege and, later, Russell took as explicandum the term 'two' in the not quite exact meaning in which it is used in everyday life and in applied mathematics; they proposed as an explicatum for it an exactly defined concept, namely, the class of pair-classes (see below the remark on (i) in § 27); other logicians have proposed other explicata for the same explicandum. Many concepts now defined in semantics are meant as explicata for concepts earlier used in everyday language or in logic. For instance, the semantical concept of truth has as its explicandum the concept of truth as used in everyday language (if applied to declarative sentences) and in all of traditional and modern logic. Further, the various interpretations of descriptions by Frege, Russell, and others, which will be discussed in §§ 7 and 8, may be regarded as so many different explications for phrases of the form 'the so-and-so'; each of these explications consists in laying down rules for the use of corresponding expressions in language systems to be constructed. The interpretation which we shall adopt following a suggestion by Frege (§8, Method IIIb) deviates deliberately from the meaning of descriptions in the ordinary language. Generally speaking, it is not required that an explicatum have, as nearly as possible, the same meaning as the explicandum; it should, however, correspond to the explicandum in such a way that it can be used instead of the latter.

The L-terms ('L-true', etc.) which we shall now introduce are likewise intended as explicata for customary, but not quite exact, concepts. ***'L-true'*** is meant as an explicatum for what Leibniz called necessary truth and Kant analytic truth. We shall indicate here briefly how this and the other L-terms can be defined. In the further discussions of this book, however, we shall not make use of the technical details of the following definitions but only of the fact that 'L-true' is defined in such a way that the requirement stated in the subsequent convention 2-1 is fulfilled. This is in accord with the purpose of this book, which is intended not so much to carry out exact analyses of exactly constructed systems as to state informally some considerations aimed at the discovery of concepts and methods suitable for semantical analysis.

[7] What is meant here by 'explicandum' and 'explicatum' seems similar to what Langford means by 'analysandum' and 'analysans'; see below, n. 42, p. 63.

We shall introduce the L-concepts with the help of the concepts of state-description and range. Some ideas of Wittgenstein[8] were the starting-point for the development of this method.[9]

A class of sentences in S_1 which contains for every atomic sentence either this sentence or its negation, but not both, and no other sentences, is called a ***state-description*** in S_1, because it obviously gives a complete description of a possible state of the universe of individuals with respect to all properties and relations expressed by predicates of the system. Thus the state-descriptions represent Leibniz' possible worlds or Wittgenstein's possible states of affairs.

It is easily possible to lay down semantical rules which determine for every sentence in S_1 whether or not it ***holds in a*** given *state-description*. That a sentence holds in a state-description means, in nontechnical terms, that it would be true if the state-description (that is, all sentences belonging to it) were true. A few examples will suffice to show the nature of these rules: (1) an atomic sentence holds in a given state-description if and only if it belongs to it; (2) $\sim \mathfrak{S}_i$ holds in a given state-description if and only if $\mathfrak{S}_i$ does not hold in it; (3) $\mathfrak{S}_i \vee \mathfrak{S}_j$ holds in a state-description if and only if either $\mathfrak{S}_i$ holds in it or $\mathfrak{S}_j$ or both; (4) $\mathfrak{S}_i \equiv \mathfrak{S}_j$ holds in a state-description if and only if either both $\mathfrak{S}_i$ and $\mathfrak{S}_j$ or neither of them hold in it; (5) a universal sentence (e.g., '$(x)(\mathrm{P}x)$') holds in a state-description if and only if all substitution instances of its scope ('Pa', 'Pb', 'Pc', etc.) hold in it. Iota-operators and lambda-operators can be eliminated (for the former, this will be shown later, see 8-2; for the latter, see the explanation of conversion in § 1). Therefore, it is sufficient to lay down a rule to the effect that any sentence containing an operator of one of these kinds holds in the same state-descriptions as the sentence resulting from the elimination of the operator.

The class of all those state-descriptions in which a given sentence $\mathfrak{S}_i$ holds is called the ***range*** of $\mathfrak{S}_i$. All the rules together, of which we have just given five examples, determine the range of any sentence in S_1; therefore, they are called *rules of ranges*. By determining the ranges, they give, together with the rules of designation for the predicates and the individual constants (e.g., 1-1 and 1-2), an *interpretation* for all sentences in S_1, since

[8] [Tractatus]; see also [I], p. 107.

[9] The method which I shall use here is similar to, but simpler than, the one I have described in [I], § 19, as procedure E. The simpler form is possible here because S_1 contains atomic sentences for all atomic propositions. The procedure to be used here seems to me the most convenient among those known at present for the semantical construction of a system of deductive logic; I have used it, furthermore, for modal logic in [Modalities] and for inductive logic, that is, the theory of logical probability or degree of confirmation in [Inductive].

to know the meaning of a sentence is to know in which of the possible cases it would be true and in which not, as Wittgenstein has pointed out.

The connection between these concepts and that of truth is as follows: There is one and only one state-description which describes the actual state of the universe; it is that which contains all true atomic sentences and the negations of those which are false. Hence it contains only true sentences; therefore, we call it the true state-description. A sentence of any form is true if and only if it holds in the true state-description. These are only incidental remarks for explanatory purposes; the definition of L-truth will not make use of the concept of truth.

The L-concepts now to be defined are meant as explicata for certain concepts which have long been used by philosophers without being defined in a satisfactory way. Our concept of L-truth is, as mentioned above, intended as an explicatum for the familiar but vague concept of logical or necessary or analytic truth as explicandum. This explicandum has sometimes been characterized as truth based on purely logical reasons, on meaning alone, independent of the contingency of facts. Now the meaning of a sentence, its interpretation, is determined by the semantical rules (the rules of designation and the rules of ranges in the method explained above). Therefore, it seems well in accord with the traditional concept which we take as explicandum, if we require of any explicatum that it fulfil the following condition:

2-1. *Convention.* A sentence $\mathfrak{S}_i$ is ***L-true*** in a semantical system S if and only if $\mathfrak{S}_i$ is true in S in such a way that its truth can be established on the basis of the semantical rules of the system S alone, without any reference to (extra-linguistic) facts.

This is not yet a definition of L-truth. It is an informal formulation of a condition which any proposed definition of L-truth must fulfil in order to be adequate as an explication for our explicandum. Thus this convention has merely an explanatory and heuristic function.

How shall we define L-truth so as to fulfil the requirement 2-1? A way is suggested by Leibniz' conception that a necessary truth must hold in all possible worlds. Since our state-descriptions represent the possible worlds, this means that a sentence is logically true if it holds in all state-descriptions. This leads to the following definition:

2-2. *Definition.* A sentence $\mathfrak{S}_i$ is ***L-true*** (in S_1) $=_{Df}$ $\mathfrak{S}_i$ holds in every state-description (in S_1).

The following consideration shows that the concept of L-truth thus defined is in accord with the convention 2-1 and hence is an adequate explicatum for logical truth. If $\mathfrak{S}_i$ holds in every state-description, then the semantical rules of ranges suffice for establishing this result. [For example, we see from the rules of ranges mentioned above that 'Pa' holds in certain state-descriptions, that '$\sim$Pa' holds in all the other state-descriptions, and that therefore the disjunction 'Pa $\vee \sim$Pa' holds in every state-description.] Therefore, the semantical rules establish also the truth of $\mathfrak{S}_i$ because, if $\mathfrak{S}_i$ holds in every state-description, then it holds also in the true state-description and hence is itself true. If, on the other hand, $\mathfrak{S}_i$ does not hold in every state-description, then there is at least one state-description in which $\mathfrak{S}_i$ does not hold. If this state-description were the true one, $\mathfrak{S}_i$ would be false. Whether this state-description is true or not depends upon the facts of the universe. Therefore, in this case, even if $\mathfrak{S}_i$ is true, it is not possible to establish its truth without reference to facts.

L-falsity is meant as an explicatum for logical or necessary falsity or self-contradiction. *L-implication* is meant as explicatum for logical implication or entailment. *L-equivalence* is intended as explicatum for mutual logical implication or entailment. The definitions are as follows:

2-3. *Definitions*

a. $\mathfrak{S}_i$ is ***L-false*** in (S_1) $=_{Df}$ $\sim\mathfrak{S}_i$ is L-true.

b. $\mathfrak{S}_i$ ***L-implies*** $\mathfrak{S}_j$ (in S_1) $=_{Df}$ the sentence $\mathfrak{S}_i \supset \mathfrak{S}_j$ is L-true.

c. $\mathfrak{S}_i$ is ***L-equivalent*** to $\mathfrak{S}_j$ (in S_1) $=_{Df}$ the sentence $\mathfrak{S}_i \equiv \mathfrak{S}_j$ is L-true.

d. $\mathfrak{S}_i$ is ***L-determinate*** (in S_1) $=_{Df}$ $\mathfrak{S}_i$ is either L-true or L-false.

The following results follow easily from these definitions, together with 2-2:

2-4. $\mathfrak{S}_i$ is L-false if and only if $\mathfrak{S}_i$ does not hold in any state-description.
2-5. $\mathfrak{S}_i$ L-implies $\mathfrak{S}_j$ if and only if $\mathfrak{S}_j$ holds in every state-description in which $\mathfrak{S}_i$ holds.
2-6. $\mathfrak{S}_i$ is L-equivalent to $\mathfrak{S}_j$ if and only if $\mathfrak{S}_i$ and $\mathfrak{S}_j$ hold in the same state-descriptions.

The condition for L-falsity stated in 2-4 means, in effect, that $\mathfrak{S}_i$ cannot possibly be true. The condition for L-implication in 2-5 means that it is not possible for $\mathfrak{S}_i$ to be true and for $\mathfrak{S}_j$ to be false. The condition for L-equivalence in 2-6 means that it is impossible for one of the two sentences to be true and the other false. Thus these results show that L-falsity, L-implication, and L-equivalence as defined by 2-3a, b, c, may indeed be regarded as adequate explicata for the explicanda mentioned earlier.

We have seen that our concept of L-truth fulfils our earlier convention 2-1. Therefore, according to the definition 2-3d, a sentence is L-determinate if and only if the semantical rules, independently of facts, suffice for establishing its truth-value, that is, either its truth or its falsity. This suggests the following definition, 2-7, as an explication for what Kant called synthetic judgments. The subsequent result, 2-8, which follows from the definition, shows that the concept defined is indeed adequate as an explicatum.

2-7. *Definition.* $\mathfrak{S}_i$ is ***L-indeterminate*** or ***factual*** (in S_1) $=_{Df}$ $\mathfrak{S}_i$ is not L-determinate.

2-8. A sentence is factual if and only if there is at least one state-description in which it holds and at least one in which it does not hold.

The concept of ***F-truth*** to be defined by 2-9a is meant as an explicatum for what is usually called factual or synthetic or contingent truth in contradistinction to logical or necessary truth. The concepts defined by 2-9b, c, d, are meant as explicata in an analogous way. The adequacy of these F-concepts as explicata follows from the adequacy of the L-concepts.

2-9. *Definitions*

a. $\mathfrak{S}_i$ is ***F-true*** (in S_1) $=_{Df}$ $\mathfrak{S}_i$ is true but not L-true.

b. $\mathfrak{S}_i$ is ***F-false*** (in S_1) $=_{Df}$ $\sim \mathfrak{S}_i$ is F-true.

c. $\mathfrak{S}_i$ ***F-implies*** $\mathfrak{S}_j$ (in S_1) $=_{Df}$ $\mathfrak{S}_i \supset \mathfrak{S}_j$ is F-true.

d. $\mathfrak{S}_i$ is ***F-equivalent*** to $\mathfrak{S}_j$ (in S_1) $=_{Df}$ $\mathfrak{S}_i \equiv \mathfrak{S}_j$ is F-true.

The following are simple consequences of these and the earlier definitions:

2-10. $\mathfrak{S}_i$ is F-false if and only if $\mathfrak{S}_i$ is false but not L-false.

2-11. $\mathfrak{S}_i$ is F-equivalent to $\mathfrak{S}_j$ if and only if $\mathfrak{S}_i$ is equivalent but not L-equivalent to $\mathfrak{S}_j$.

As an example of F-truth, consider the sentence 'Bs'. We found earlier with the help of a rule of truth and rules of designation, that 'Bs' is true if and only if Scott is a biped (1-4). This result does not tell us whether 'Bs' is true or not; it merely states a sufficient and necessary condition for the truth of the sentence 'Bs'. This is all we can learn about 'Bs' from the semantical rules alone. If we want to determine the truth-value of 'Bs', we have to go beyond the mere semantical analysis to the observation of facts. We see from 1-4 which facts are relevant: we must look at the thing Walter Scott and see whether it is a biped. Observation shows that this is the case. Therefore, 'Bs' is true. Since the semantical rules do not suffice for establishing its truth, it is not L-true; hence it is F-true.

§ 3. Equivalence and L-Equivalence

The symbol '$\equiv$', customarily used between sentences, is used here also between designators of other kinds, especially between predicators and between individual expressions. 'P $\equiv$ Q' is to mean the same as '(x) (Px $\equiv$ Qx)'. 'a $\equiv$ b' is used, instead of the customary 'a = b', as an identity sentence, saying that a is the same individual as b. Then the concepts of equivalence and L-equivalence, previously applied to sentences only, are defined for designators of any kind; these two concepts are fundamental in our method. Two designators are said to be *equivalent* if the $\equiv$-sentence connecting them is true; they are said to be *L-equivalent* if this sentence is L-true. It follows that 'P' and 'Q' are equivalent if they hold for the same individuals. And 'a' and 'b' are equivalent if a is the same individual as b.

We have defined the terms 'equivalent' and 'L-equivalent' so far only for sentences (1-8 and 2-3c). Now we shall extend their use so as to make them applicable to all kinds of designators, especially also to predicators and individual expressions. Extended in this way, the two concepts will become the fundamental concepts in the method of semantical analysis to be proposed here.

We begin by extending the use of the symbol '$\equiv$'. It is customary as a connective between sentences. We shall use it in our systems between two designators of any kind, but only if both designators are of the same type. This use is introduced by the following rules of abbreviation. If the extended use of '$\equiv$' is taken as primitive, then suitable rules of ranges are to be laid down which lead to the same results (for example, the result that 'P $\equiv$ Q' has the same range as, and hence is L-equivalent to, '(x)(Px $\equiv$ Qx)'). The reasons for choosing just these interpretations for '$\equiv$' with the various kinds of designators will soon become apparent.

The first rule introduces '$\equiv$' between predicators:

3-1. *Abbreviation*

a. Let $\mathfrak{A}_i$ and $\mathfrak{A}_j$ be two *predicators* of the same degree n in S_1.
$\mathfrak{A}_i \equiv \mathfrak{A}_j$ for $(x_1)(x_2)\ ..\ (x_n)[\mathfrak{A}_i x_1 x_2\ ..\ x_n \equiv \mathfrak{A}_j x_1 x_2\ ..\ x_n]$.

b. Hence for degree one:
$\mathfrak{A}_i \equiv \mathfrak{A}_j$ for $(x)[\mathfrak{A}_i x \equiv \mathfrak{A}_j x]$.

We shall use in S_1 the connective '$\bullet$' also between predicators, but, for the sake of a convenient notation, in a way different from the use of '$\equiv$' just introduced. The resulting expression (e.g., 'P$\bullet$Q') is here taken as a predicator, not as a sentence as in the case of '$\equiv$' (e.g., 'P $\equiv$ Q'). We define it for degree one:

3-2. *Abbreviation.* Let $\mathfrak{A}_i$ and $\mathfrak{A}_j$ be two predicators of degree one in S_1. $\mathfrak{A}_i \bullet \mathfrak{A}_j$ for $(\lambda x)[\mathfrak{A}_i x \bullet \mathfrak{A}_j x]$.

Thus, for example, '$F \bullet B$' is short for '$(\lambda x)[Fx \bullet Bx]$', and hence is an expression for the property of being a featherless biped.

Furthermore, we introduce '$\equiv$' as a primitive sign of identity of individuals instead of the customary '$=$'. For this purpose we lay down the following rule:

3-3. *Rule of truth.* If $\mathfrak{A}_i$ is an individual expression in S_1 for the individual x and $\mathfrak{A}_j$ for y, then $\mathfrak{A}_i \equiv \mathfrak{A}_j$ is true if and only if x is the same individual as y.

[If S is an extensional system containing, in distinction to S_1, a predicator variable 'f', then we can achieve the same result as 3-3 by defining $\mathfrak{A}_i \equiv \mathfrak{A}_j$, in a way similar to Russell's, as short for $(f)[f(\mathfrak{A}_i) \equiv f(\mathfrak{A}_j)]$.]

If a system S contains, in distinction to S_1, *functors* also, then '$\equiv$' can be defined for them in a way similar to the above definition for predicators. The method may be indicated briefly by stating the definition for the simplest type, namely, functors for singulary functions from individuals to individuals; the definitions for other types are analogous. This definition will not be used in our further discussions.

3-4. *Abbreviation.* For functors $\mathfrak{A}_i$ and $\mathfrak{A}_j$ in S:

$$\mathfrak{A}_i \equiv \mathfrak{A}_j \text{ for } (x)[\mathfrak{A}_i x \equiv \mathfrak{A}_j x].$$

[Note that here on the right-hand side the sign '$\equiv$' stands, not between sentential matrices, as in 3-1b, but between full expressions of functors, which are for this type individual expressions.]

Now we shall define 'equivalent', 'L-equivalent', and 'F-equivalent' in a general way for all kinds of designators.

3-5. *Definitions.* Let $\mathfrak{A}_i$ and $\mathfrak{A}_j$ be two designators of the same type in S_1.

a. $\mathfrak{A}_i$ is ***equivalent*** to $\mathfrak{A}_j$ in (S_1) $=_{Df}$ the sentence $\mathfrak{A}_i \equiv \mathfrak{A}_j$ is true (in S_1).

b. $\mathfrak{A}_i$ is ***L-equivalent*** to $\mathfrak{A}_j$ (in S_1) $=_{Df}$ $\mathfrak{A}_i \equiv \mathfrak{A}_j$ is L-true (in S_1).

c. $\mathfrak{A}_i$ is ***F-equivalent*** to $\mathfrak{A}_j$ (in S_1) $=_{Df}$ $\mathfrak{A}_i \equiv \mathfrak{A}_j$ is F-true (in S_1).

Now let us see what the concepts just defined mean for the various kinds of designators. We begin with *predicators*. Let 'P' and 'Q' be two predicators of degree one in S_1. According to 3-5a, they are equivalent if and only if '$P \equiv Q$' is true, hence, according to 3-1b, if and only if '$(x)[Px \equiv Qx]$' is true, hence if 'P' holds for the same individuals as 'Q'. The result is analogous for two predicators of any degree n, say 'R' and 'R′'. They are equivalent, according to 3-5a and 3-1a, if and only if

'$(x_1) \ldots (x_n)[\mathrm{R}x_1 \ldots x_n \equiv \mathrm{R}'x_1 \ldots x_n]$' is true, hence if the two predicators hold for the same sequences (of length n) of individuals.

To give an example, let us assume the following as a biological fact:

3-6. *Assumption.* All human beings are featherless bipeds and vice versa.

Then the following holds:

3-7. The sentence '$(x)[\mathrm{H}x \equiv (\mathrm{F}\bullet\mathrm{B})x]$' is true (in S_1), but not L-true, hence F-true.

According to 3-1b, the sentence just mentioned can be abbreviated by '$\mathrm{H} \equiv \mathrm{F}\bullet\mathrm{B}$'. Hence, 3-5 yields:

3-8. The predicators 'H' and '$\mathrm{F}\bullet\mathrm{B}$' are equivalent (in S_1), but not L-equivalent, hence F-equivalent.

On the other hand, the truth of the sentence '$(x)[\mathrm{H}x \equiv \mathrm{RA}x]$' can be established without referring to facts by merely using the semantical rules of S_1, especially 1-2 (see the remark following this rule) and the truth rules for the universal quantifier and for '$\equiv$'. Therefore:

3-9. '$(x)[\mathrm{H}x \equiv \mathrm{RA}x]$' is L-true.

According to 3-1b, the sentence just mentioned can be abbreviated by '$\mathrm{H} \equiv \mathrm{RA}$':

3-10. '$\mathrm{H} \equiv \mathrm{RA}$' is L-true.

Hence, 3-5b yields:

3-11. The predicators 'H' and 'RA' are L-equivalent (in S_1).

Now let us apply our definitions to *individual expressions*. The following result is obtained from 3-3 and 3-5a:

3-12. Individual expressions are equivalent if and only if they are expressions for the same individual.

Examples for L-equivalence and F-equivalence of individual expressions will be given later (§ 9).

A consideration of these results for predicators and individual expressions shows the following: If 'P' and 'Q' are equivalent predicators, then 'Pa' and 'Qa' are either both true or both false and hence, in any case, equivalent; the same holds for 'Pb' and 'Qb', etc. Furthermore, if 'a' and 'b' are equivalent, then 'Pa' and 'Pb' are either both true or both false and hence, in any case, equivalent; the same holds for 'Qa' and 'Qb', etc. An analogous result for functors follows from rules like 3-4. It can be shown that the following two theorems hold generally for our systems S_1, S_2, and S_3, and likewise for any similar systems, including those containing functors, provided that definitions analogous to those given above are laid down.

3-13. If two designator signs are equivalent, then any two sentences of simplest form (in S_1: atomic form) which are alike except for the occurrence of the two designator signs are likewise equivalent.

3-14. If two designators (which may be compound expressions) are L-equivalent, then any two sentences (of any form whatever) which are alike except for the occurrence of the two designators are likewise L-equivalent.

These two results show that our choice of the interpretation for the extended uses of '≡' and of the definition for the extended use of the terms 'equivalent' and 'L-equivalent' was not arbitrary. In fact, the choice was made with the intention of reaching these results. In particular, the first result 3-13, in its application to individual expressions, may be regarded as supplying a justification for the use of '≡' as a sign of identity, which might at first perhaps appear strange.

On the basis of equivalence and L-equivalence for designators we define the following two concepts:

3-15. *Definitions.* Let $\mathfrak{A}_i$ be a designator (in S_1).

a. The ***equivalence class*** of $\mathfrak{A}_i =_{Df}$ the class of those expressions (in S_1) which are equivalent to $\mathfrak{A}_i$.

b. The ***L-equivalence class*** of $\mathfrak{A}_i =_{Df}$ the class of those expressions (in S_1) which are L-equivalent to $\mathfrak{A}_i$.

It is easily seen that $\mathfrak{A}_i$ itself belongs to both classes, that the L-equivalence class is a subclass of the equivalence class, and that both classes contain only designators of the same type as $\mathfrak{A}_i$.

§ 4. Classes and Properties

It is customary to regard two *classes*, say those corresponding to the predicators 'P' and 'Q', as identical if they have the same elements, in other words, if 'P' and 'Q' are equivalent. We regard the two *properties* P and Q as identical if 'P' and 'Q' are, moreover, L-equivalent. By the *intension* of the predicator 'P' we mean the property P; by its *extension* we mean the corresponding class. It follows that two predicators have the same extension if they are equivalent, and the same intension if they are L-equivalent. The term 'property' is to be understood in an objective, physical sense, not in a subjective, mental sense; the same holds for terms like 'concept', 'intension', etc. The use of these and related terms does not involve a hypostatization.

In analyzing the meaning of an adjective, e.g., 'human', or a corresponding predicator in a symbolic language, e.g., 'H', it is customary to speak of two entities—on the one hand, the property of being human or, as we shall write for short, the property Human; on the other hand, the class of human beings, or the class Human.[10]

The metalanguage M must contain certain translations of the sentences of the object languages to be dealt with in M. The translation can often be formulated in different ways. Take as an example an atomic sentence in S_1, say 'Hs'. Its simple, straightforward translation into M is as follows, according to our rules of designation for 'H' and 's' (1-2 and 1-1):

4-1. 'Scott is human'.

There are two other translations of 'Hs' which in a sense are more explicit by using the terms 'property' or 'class' but which have the same logical content as 4-1:

4-2. 'Scott has the property Human'.

4-3. 'Scott belongs to (is an element of) the class Human'.

As another example, take the sentence '$(x)[Hx \supset Bx]$'. Here, likewise, there is a direct translation (4-4) and two more explicit ones with 'property' (4-5) or 'class' (4-6):

4-4. 'For every x, if x is human, then x is a biped'.

4-5. 'The property Human implies (materially) the property Biped'.

4-6. 'The class Human is a subclass of the class Biped'.

In these examples the terms 'property' and 'class' seem unnecessary, since there are forms which avoid those terms (4-1 and 4-4). Thus the important question may be raised as to whether semantics could not do entirely without those terms. However, we shall first accept them, so to speak, uncritically, endeavoring merely to make their customary use more exact and consistent. Later only shall we come back to the question mentioned; it will then be shown how the apparent multiplicity of entities which seems to be introduced by the admission of these and other terms can be reduced (§§ 33 f.). Thus our present acceptance of the two more explicit forms of translation is merely an introduction of two ways of speaking; it does by no means imply the recognition of two separate kinds of entities—properties, on the one hand; classes, on the other.

[10] Since a brief formulation seems desirable and since phrases of the form 'the property human' and 'the class human' are contrary to English grammar and sometimes even ambiguous, I have used in earlier publications (see [I], p. 237) double quotation marks, e.g., 'the property "human" '. However, this use of quotation marks differs from their normal use. Therefore, I prefer now the method of *capitalizing;* I shall use it not only in connection with 'property' and 'class' but likewise with other words designating kinds of entities, e.g., 'relation', 'function', 'concept', 'individual', 'individual concept', and the like. In connection with nouns instead of adjectives I often use also the customary form with 'of', e.g., I write either 'the concept of equivalence' or 'the concept Equivalence'.

The above examples seem to show a certain parallelism between the two modes of speech, the one in terms of 'property' and the other in terms of 'class'. However, there is one fundamental difference, leaving aside the inessential, merely idiomatic difference that in the one case the connecting phrase is 'has' or 'possesses', while in the other it is 'belongs to' or 'is an element of'. The fundamental difference is in the *condition of identity*. Classes are usually taken as identical if they have the same elements. Thus, for example, on the basis of our earlier assumption (3-6), the class Human has the same elements as the class Featherless Biped. Therefore:

4-7. The class Human is the same as the class Featherless Biped.

Under what conditions properties are usually regarded as identical is less clear. It seems natural, and sufficiently in agreement with the vague customary usage, to regard properties as identical if it can be shown by logical means alone, without reference to facts, that whatever has the one property has the other and vice versa; in other words, if the equivalence sentence is not only true but L-true. Thus with respect to our earlier examples (3-7 and 3-9) the following holds:

4-8. The property Human is not the same as the property Featherless Biped.

4-9. The property Human is the same as the property Rational Animal.

It is easily seen, on the basis of our definitions in the preceding section (3-1b and 3-5a, b) that the identity conditions stated above can be formulated in the following way with respect to predicators (of degree one):

4-10. Classes are identical if and only if predicators for them are equivalent.

4-11. Properties are identical if and only if predicators for them are L-equivalent.

Now we shall introduce the terms ***'extension'*** and ***'intension'*** with respect to predicators. If two predicators apply to the same individuals—in other words, if they are equivalent—it is sometimes said that they are coextensive or that they have the same extension (in one of the various customary uses of this term). The use of 'intension' varies still more than that of 'extension'. It seems in agreement with at least one of the customary usages to speak of the same intension in the case of L-equivalence. Thus we lay down the following two *conventions:*

4-12. Two predicators ***have the same extension*** if and only if they are equivalent.

4-13. Two predicators ***have the same intension*** if and only if they are L-equivalent.

These conventions determine only the use of the phrases 'have the same extension' and 'have the same intension'. For many purposes this is sufficient. If, however, we wish to go further and to speak of something as the extension of a given predicator, and of something else as its intension, then these conventions do not suffice; but they help us by narrowing the choice of suitable entities. The first convention means that we may take as extensions of predicators only something which equivalent predicators have in common. According to 4-10, this condition is fulfilled by the corresponding classes. The second convention means that we may take as intensions of predicators only something which L-equivalent predicators have in common. According to 4-11, this condition is fulfilled by the corresponding properties. This suggests the following conception of the extension and the intension of predicators:

4-14. The *extension of a predicator* (of degree one) is the corresponding class.

4-15. The *intension of a predicator* (of degree one) is the corresponding property.

This seems sufficiently in agreement with customary usage. If this is applied to the predicator 'H' in S_1, we obtain:

4-16. The extension of 'H' is the class Human.

4-17. The intension of 'H' is the property Human.

Both results hold also for the predicator '$(\lambda x)(Hx)$', which is L-equivalent to 'H' in S_1.

It is obvious that there are many other ways for choosing entities as extensions and intensions of predicators (of degree one) so as to satisfy our conventions (4-12 and 4-13). One alternative is as follows: It is possible to take as the extension of a predicator its equivalence class (3-15a) and as its intension its L-equivalence class (3-15b). This conception seems less natural than the one we have chosen (4-14, 4-15), because it leads to linguistic instead of to extra-linguistic entities. On the other hand, this conception of intensions has the advantage that it is possible in an extensional metalanguage; this will be explained later. (Compare definitions by Russell and Quine mentioned below, at the end of § 33.)

It may perhaps be useful, in order to avoid misunderstandings, to add some informal remarks concerning the use in this book of the term ***'property'***. This term will be used as synonymous with words like 'quality', 'character', 'characteristic', and the like in their ordinary use. It is to be understood in a very wide sense, including whatever can be said meaningfully, no matter whether truly or falsely, about any individual. The term is used here not only for qualitative properties in the narrower sense (for example, the properties Blue, Hot, Hard, and the like) but also for quantitative properties (for example, the property Weighing Five Pounds), for relational properties (e.g., the property Uncle Of Somebody), for spatiotemporal properties (e.g., the property North Of Chicago), and others. It is important to note what is *not* meant here by the term 'property'. First, it does not refer to linguistic expressions; to the symbol 'H' and the corresponding word 'human' we apply the term 'predicator', not 'property'; by a property we mean rather what is expressed by a predicator (of degree one). Second, the properties of things are not meant as something mental, say images or sense-data, but as something physical that the things have, a side or aspect or component or character of the things. If an observer sees that this table is red, then the table has the character Red and the observer has the corresponding character Red-Seeing. By the property Red we mean the first, not the second; we mean that physical character of the thing which the physicist explains as a certain disposition to selective reflection, not that psychological character of the observer which the physiologist explains as a certain disposition to a specific reaction by the sensory part of the nervous system.

Suppose we understand some predicators in a given language; that is to say, we know which properties they express. Suppose, further, that we have experienced each of these properties; that is to say, we have, for each of them, found some things which, according to our observations, have that property. We can form compound predicators out of the given predicators with the help of logical particles. Then we understand a compound predicator because its meaning is determined by the meanings of the component predicators and the logical structure of the compound expression. It is important to notice that our understanding of a compound predicator is no longer dependent upon observations of any things to which it applies, that is, any things which have the complex property expressed by it.

In order to construct examples, suppose that the system S_1 contains not only the predicator 'H' for the property Human, but also the predicator 'T' for the property Twenty Feet High. Then we can, for example, form the following compound predicators (provided we permit the use of '$\sim$' and '$\vee$' in predicators in analogy to the use of '$\bullet$' introduced by 3-2): '$\sim$H' expresses the property Non-Human, 'H $\vee$ T' the property Human

Or Twenty Feet High, and 'H•T' the property Human And Twenty Feet High. We know things which exemplify the first of these three properties, and likewise some for the second. But we have never seen any things that exemplify the predicator 'H•T', and there are presumably no things of this kind in the world. Nevertheless, 'H•T' is not meaningless. Since it is a well-formed predicator (of degree one), it expresses a property, although this property does not apply anywhere. We shall say of both the predicator and the property that they are *empty*. One can understand 'H•T' just as clearly as the other compound predicators; and one may indeed understand this or any other compound predicator before he knows whether and, if so, where it is exemplified. The understanding of a compound predicator is based upon the understanding of the component predicators. Exemplification in experience is required only for primary predicators, with the help of which the others are interpreted.

Now consider the predicator 'H• ∼H'. No factual knowledge is needed for recognizing that this predicator cannot possibly be exemplified. Nevertheless, this expression is not meaningless. It is a well-formed predicator; it expresses the property Human And Non-Human.[11] We shall say of both the predicator and the property that they are *L-empty* (logically empty). [There is only one L-empty property, although there are many empty properties. If 'P' and 'Q' are any two L-empty predicators, then '$P \equiv Q$', that is, '$(x)(Px \equiv Qx)$' (3-1b), is L-true; therefore, 'P' and 'Q' are L-equivalent (3-5b); hence they express the same property (4-11).]

The use of the term '*relation*' in this book is analogous to that of the term 'property' just explained. A relation is meant neither as a mental entity nor as an expression but rather as something that is expressed by certain designators, namely, predicators of degree two or more, and that may hold objectively for two or more things.

The term '***concept***' will be used here as a common designation for properties, relations, and similar entities (including individual concepts, to be explained in § 9, and functions, but not propositions). For this term it is especially important to stress the fact that it is not to be understood in a mental sense, that is, as referring to a process of imagining, thinking, conceiving, or the like, but rather to something objective that is found in nature and that is expressed in language by a designator of nonsentential form. (This does not, of course, preclude the possibility that a concept—for example, a property objectively possessed by a given thing—may be subjectively perceived, compared, thought about, etc.)

[11] Compare Bennett and Baylis, [Logic], sec. 3.4: "The existence of self-inconsistent concepts."

The preceding remarks are meant merely as an informal terminological clarification. They should by no means be regarded as an attempt toward a solution of the old controversial problem of the universals. The traditional discussions concerning this problem are, in my view, a rather heterogeneous mixture of different components, among them logical statements, psychological statements, and pseudo-statements, that is, expressions which are erroneously regarded as statements but do not have cognitive content, although they may have noncognitive—for instance, emotive—meaning components. My remarks on the interpretation intended for the term 'property' are admittedly rather vague, chiefly because of a lack of a clear and generally accepted terminology about matters of this kind. Nevertheless, I hope they will give sufficiently clear indications for all practical purposes and, above all, may help to avoid certain typical misunderstandings.

I wish to emphasize the fact that the discussions in this book about properties, and similarly about relations, concepts in general, propositions, etc., do not involve a hypostatization. As I understand it, a hypostatization or substantialization or reification consists in mistaking as things entities which are not things. Examples of hypostatizations of properties (or ideas, universals, or the like) in this sense are such formulations as 'the ideas have an independent subsistence', 'they reside in a super-heavenly place', 'they were in the mind of God before they became manifested in things', and the like, provided that these formulations are meant literally and not merely as poetical metaphors. (We leave aside here the historical question of whether these hypostatizations are to be attributed to Plato himself or rather to his interpreters.) These formulations, if taken literally, are pseudo-statements, devoid of cognitive content, and therefore neither true nor false. Whatever is said in this book about properties may be wrong, but it has at least cognitive content. This follows from the fact that our statements belong to, or can be translated into, the general language of science. We use the term 'property' in that sense in which it is used by scientists in statements of the following form: 'These two bodies have the same chemical properties, but there are certain physical properties in which they differ'; 'Let us express the property . . . , which is exemplified by the one but not by the other of these two bodies, by 'P' '.

The term ***'entity'*** is frequently used in this book. I am aware of the metaphysical connotations associated with it, but I hope that the reader will be able to leave them aside and to take the word in the simple sense in which it is meant here, as a common designation for properties, propositions, and other intensions, on the one hand, and for classes, individuals, and other extensions, on the other. It seems to me that there is no other suitable term in English with this very wide range.

§ 5. Extensions and Intensions

In analogy to the case of predicators, we shall say of two designators of any kind that they have the same *extension* if they are equivalent, and that they have the same *intension* if they are L-equivalent. In later sections we shall discuss the problem of finding suitable entities which might be taken as extensions and intensions in accordance with these identity conditions. If two predicators, say 'P' and 'Q', are equivalent or L-equivalent in a system *S*, we say also that the properties P and Q are equivalent or L-equivalent, respectively; and analogously with designators of other kinds and their intensions.

In the preceding section we introduced the terms 'extension' and 'intension' with respect to predicators only, in agreement with traditional usage. Now we shall extend the use of these terms, applying them to other types of designators in an analogous way.

In the case of predicators, we have taken equivalence as the condition for identity of extension, and L-equivalence for identity of intension (4-12 and 4-13). Earlier (§ 3), we saw how the semantical concepts of equivalence and L-equivalence can be applied to the various types of designators. Thus it seems natural to take the same conditions as defining identity of extension or intension with respect to designators in general. This leads to the following definitions; 4-12 and 4-13 are now regarded simply as special cases hereof.

5-1. *Definition.* Two designators ***have the same extension*** (in S_1) $=_{Df}$ they are equivalent (in S_1).

5-2. *Definition.* Two designators ***have the same intension*** (in S_1) $=_{Df}$ they are L-equivalent (in S_1).

Note that the terms 'extension' and 'intension' have not been defined hereby, but only the phrases 'have the same extension' and 'have the same intension'. In order to speak about extensions and intensions themselves, we have to look for entities, or at least for phrases apparently referring to entities, which can be assigned to designators in accordance with these definitions. In the case of predicators, we found classes and properties as such entities. We shall see later how suitable entities can be chosen for sentences and individual expressions.

The introduction into the metalanguage M of expressions for additional kinds of entities is always a precarious step that must be taken with caution and with careful consideration of the consequences. We shall discuss the problem involved in the introduction of extensions and intensions for designators later (§§ 33 ff.). Here it may be noted that the phrases 'have the same extension' and 'have the same intension', although apparently referring to certain entities as extensions and intensions, are, in fact, entirely free of the problematic nature of the terms 'extension' and 'intension'; for those phrases are based by the above definitions on the terms 'equivalent' and 'L-equivalent', and these go back (by 3-5) to the terms 'true' and 'L-true', which can be defined for the system S_1 in an exact way, as explained earlier.

It is often convenient to apply the *term 'equivalent'*, and perhaps also the term 'L-equivalent', not only to designators but likewise *to the intensions* of these designators; thus not only to predicators (e.g., 'the predicators 'H' and 'F•B' are equivalent in S_1') but also to properties and relations (e.g., 'the property Human and the property Featherless Biped are equivalent'); and analogously not only to sentences but also to propositions. This transferred use cannot lead to any actual ambiguity or confusion, for two reasons: (1) The context always makes clear whether the term 'equivalent' is meant in the original or in the transferred sense; the former is the case whenever the term is applied to expressions in a language system, the latter whenever it is applied to intensions, hence to extralinguistic entities. (2) In the original use the term is accompanied by a reference to a language system (e.g., 'equivalent in S_1'; however, this holds only for the complete formulation; in practice we often omit the reference if the context makes clear which language system is meant); the transferred use is not so accompanied (e.g., it makes no sense to say 'these two properties are equivalent in S_1'). Two designators may be equivalent in one language and not in another, because they may have other meanings in the second language; thus the equivalence of designators is dependent upon the language, as all semantical concepts are. On the other hand, the equivalence of two properties is not dependent upon language; it is a nonsemantical and, moreover, a nonlinguistic concept (e.g., it is a biological, not a linguistic, fact that the property Human and the property Featherless Biped are equivalent). The term 'equivalent' in the transferred use still belongs to the metalanguage M; not, however, to the semantical part of M but to what we might call the object part, that is, that part of M which contains the translations of the sentences and other expressions of the object languages.[12] The application of the terms 'equivalent' and

[12] Previously, I called terms of this kind, which are transferred from semantics to extralinguistic entities, *absolute terms* ([I], § 17), in order to indicate that in their new use the terms are no longer relative to a language. However, I now prefer to avoid the word 'absolute' because some readers were puzzled by it and suspected behind it some sort of metaphysical absolutism.

'L-equivalent' to intensions of designators, if these designators are equivalent or L-equivalent in the original semantical sense, leads, in combination with the identity conditions expressed in 5-1 and 5-2, to the following results:

5-3. If two designators are equivalent (in S_1), then we say that their extensions are identical and that their intensions are equivalent.

5-4. If two designators are L-equivalent (in S_1), then we say that their intensions are L-equivalent (or identical).

Because of 5-3, '≡' may be regarded as a sign both for the identity of extensions and for the equivalence of intensions; in particular, if it stands between predicators of degree one (as in 'H ≡ F•B'), it is a sign of identity of classes and a sign of equivalence of properties.

Examples. We found earlier that the predicators 'H' and 'F•B' are equivalent but not L-equivalent (3-8), and that 'H' and 'RA' are L-equivalent (3-11). If we apply here the above two definitions, we obtain the following formulations with transferred terms, in addition to the earlier formulations in terms of identity (4-7, 4-8, and 4-9):

5-5. The property Human is equivalent to the property Featherless Biped.

5-6. The property Human is not L-equivalent to the property Featherless Biped.

5-7. The property Human is L-equivalent to the property Rational Animal.

Of these three formulations, only the first is actually useful in M; the other two serve only as preparation for analogous formulations in another metalanguage M′ (§ 34). [It may be remarked incidentally that the terms 'equivalent' and 'L-equivalent' in their transferred, nonsemantical use, which are here applied to intensions, could also be applied to extensions. However, equivalence of extensions would be the same as identity of extensions and hence would not be useful. And to speak of L-equivalence of extensions would even be dangerous because it would lead to the same consequences that we shall later find for sentences like 42-6A.]

§ 6. Extensions and Intensions of Sentences

We take as the extension of a sentence its truth-value, and as its intension the proposition expressed by it. This is in accord with the identity conditions for extensions and for intensions stated in the preceding section. Propositions are here regarded as objective, nonmental, extra-linguistic entities. It is shown that this conception is applicable also in the case of false sentences.

Now let us see whether we can find entities which may be taken as extensions and intensions of sentences in accordance with our definitions for the identity of extensions (5-1) and of intensions (5-2).

According to 5-1, we must take as extensions of sentences something that equivalent sentences have in common. The most natural choice seems the truth-values:

6-1. The *extension of a sentence* is its truth-value.

At first glance, it may perhaps seem strange to call a truth-value an extension, and perhaps there may be a feeling even against saying that equivalent sentences have the same extension. The term 'extension' seems natural enough in the case of predicators; we easily visualize the domain of individuals as an area and the class of individuals to which a certain predicator applies (e.g., the class Biped for the predicator 'B') as a subarea which extends over a smaller or larger part of the whole. But one might say that in the case of a truth-value there is nothing extended. However, a closer inspection may remove the impression of strangeness. It has become customary to use the term 'extensional' for truth-functional connections, i.e., for connections such that the truth-value of the full sentence is a function of the truth-values of the components. And there is, indeed, a strong analogy between truth-values of sentences and extensions of predicators. This can be seen as follows: A predicator of degree n is characterized by the fact that we must attach to it n argument expressions in order to form a sentence. Therefore, a sentence might be regarded as a predicator of degree zero. Let $\mathfrak{A}_i$ and $\mathfrak{A}_j$ be any predicators of degree n ($n \geqq 1$); then (according to 4-12, 3-5, and 3-1a) $\mathfrak{A}_i$ and $\mathfrak{A}_j$ have the same extension if and only if $(x_1)(x_2) \ldots (x_n)[\mathfrak{A}_i x_1 x_2 \ldots x_n \equiv \mathfrak{A}_j x_1 x_2 \ldots x_n]$ is true. If we stipulate that this, which applies originally only to $n \geqq 1$, is to be applied analogously to sentences as predicators of degree zero, we find that two sentences, $\mathfrak{S}_i$ and $\mathfrak{S}_j$, have the same extension if and only if $\mathfrak{S}_i \equiv \mathfrak{S}_j$ is true, hence if and only if $\mathfrak{S}_i$ and $\mathfrak{S}_j$ are equivalent. Thus we are led back to 5-1 as applied to sentences; and then it seems natural to take the truth-values as extensions. [For the time being we may leave aside the question of what kind of entities these truth-values are, which are here proposed as extensions. This problem will be discussed later (in § 23).]

Now we have to decide what entities to take as intensions of sentences. It is often said that a (declarative) sentence expresses a proposition. We accept this use of the word 'proposition'; that is to say, we do not use this word for sentences or for sentences together with their meaning but for those entities which themselves are extra-linguistic but which, if they find expression in a language, are expressed by (declarative) sentences.[13] Those authors who use the term 'proposition' in this sense are often not quite clear as to the condition under which two sentences express the same proposition. We decide to take L-equivalence as this condition.[14] Thus, for example, we say that the sentences '$\sim$(Pa•Qb)' and '$\sim$Pa V $\sim$Qb' express the same proposition. This seems sufficiently in agreement with the usage of many logicians. Since we took L-equivalence as the condition of identity for intensions (5-2), we may regard propositions as intensions:

6-2. The *intension of a sentence* is the proposition expressed by it.

Examples:

6-3. The extension of the sentence 'Hs' (in S_1) is the truth-value that Scott is human,[15] which happens to be the truth.

6-4. The intension of the sentence 'Hs' is the proposition that Scott is human.[15]

Some remarks may help to clarify the sense in which we intend to use the term '***proposition***'. Like the term 'property' (§ 4), it is used neither for a linguistic expression nor for a subjective, mental occurrence, but rather for something objective that may or may not be exemplified in nature. [We might say that propositions are, like properties, of a conceptual nature. But it may be better to avoid this formulation, because it might lead to a subjectivistic misinterpretation, if the fact is overlooked that we use the term 'concept' in an objective sense (see § 4).] We apply the term 'proposition' to any entities of a certain logical type, namely, those that may be expressed by (declarative) sentences in a language. By the property Black we mean something that a thing may or may not have and that this table actually has. Analogously, by the proposition that this table is black we mean something that actually is the case with this table, something that is exemplified by the fact of the table's being as it is.

[13] On the necessity of distinguishing clearly between the two meanings of the term 'proposition', compare [I], pp. 235 f.

[14] Compare [I], p. 92.

[15] In analogy to 'the property Human' and 'the class Human' we might write here 'the proposition Scott-Is-Human' and 'the truth-value Scott-Is-Human'. However, this would become rather awkward for longer sentences. Therefore, we shall instead insert 'that' after 'proposition', thus coming back to ordinary usage. For the sake of analogy, we shall likewise write 'the truth-value that . . .', although it deviates from ordinary usage; we cannot use the more idiomatic form 'the truth-value of the proposition that . . .' because in 6-3 we wish to speak only about the sentence and its extension, the truth-value, not about its intension, the proposition.

(This simple explanation is possible only in the case of a true proposition; the problem of false propositions will soon be discussed.)

The question of whether *facts* are propositions of a certain kind or entities of a different nature is controversial. Ducasse[16] identifies facts with true propositions. Bennett and Baylis[17] say that propositions are true or false; on the other hand, "facts themselves are neither true nor false, but just *are*". The question is, to a certain extent, a terminological one and hence to be settled by convention. Since the term 'fact' in its ordinary use is rather vague and ambiguous, there is some freedom of choice left as to how to turn it into an exact technical term, in other words, how to explicate it. I am inclined to think, like Ducasse, that it would not deviate too much from customary usage if we were to explicate the term 'fact' as referring to a certain kind of proposition (in our objective sense of this word). What properties must a proposition have to be a fact in this sense? First, it must, of course, be true; second, it must be contingent (or factual); thus it must be F-true. I think that still another requirement should be added: The proposition must be specific or complete in a certain sense; but I am not sure what degree of completeness should be required. An example may illustrate the problem. The proposition that this thing (a piece of paper I have before me) is blue is a true proposition; in other words, this thing has the property Blue. But the property Blue has a wide range; it is not specific but includes many different shades of blue, say Blue_1, Blue_2, etc. This thing, on the other hand, or, more exactly speaking, a specified position c on its surface at the present moment has only one of these shades, say Blue_5. Let p be the proposition that c is blue, and q the more specific proposition that c is blue_5. It is the truth of q that makes p true. Therefore, the nonspecific proposition p should perhaps not be regarded as a fact. Whether q should be so regarded remains doubtful; q is completely specific in one respect, concerning the color, but it does not specify the other properties of the given thing. Should we require complete specificity with respect to all properties of the thing or things involved, and also with respect to all relations among the given things, or perhaps even with respect to all relations between the given things and all other things? It seems somewhat arbitrary to draw a line at any of these points. If we do not stop at some point but go the whole way, then we arrive at the strongest F-true proposition p_T, which is the conjunction of

[16] C. J. Ducasse, "Propositions, Opinions, Sentences, and Facts", *Journal of Philosophy*, XXXVII (1940), 701–11; see also his reply to some objections (*ibid.*, XXXIX [1942], 132–36).

[17] [Logic], p. 49.

all true propositions[18] and hence L-implies every true proposition. If we require of a fact this maximum degree of completeness (short of L-falsity), then there is only one fact, the totality of the actual world, past, present, and future. We indicate here these various possibilities for choosing an explicatum for the concept of fact without making a decision. We shall not take the term 'fact' as a technical term but shall use it only in informal explanations; thus, for example, we have said (§ 2) that the truth-value of a sentence which is not L-determinate is dependent upon the facts.

The greatest difficulty in the task of explicating the concept of proposition is involved in the case of a false sentence. Since this piece of paper c is, in fact, blue, sentences like 'c is not blue' or 'c is red' are false. They cannot be regarded as meaningless, because we understand their meaning before we know whether they are true or false. Therefore, these sentences, too, express propositions. On the other hand, these propositions cannot have the same relation to facts as the proposition expressed by the true sentence 'c is blue'. While the latter proposition is exemplified by a fact, the former ones are not. What, then, are these false propositions? Are there any entities of which we can say that they are expressed by those false sentences, but for which we cannot point out any exemplifying facts?

Russell has given a thorough discussion of the problems here involved. He likewise decides to use the term 'proposition' for what is expressed by a sentence, in other words, for the signification of a sentence, provided that an entity of this kind can be found. But he despairs of finding an entity of this kind in the objective, factual realm. He argues as follows: "Since a significant sentence *may* be false, it is clear that the signification of a sentence cannot be the fact that makes it true (or false). It must, therefore, be something in the person who believes the sentence, not in the object to which the sentence refers."[19] "Propositions . . . are to be defined as psychological and physiological occurrences of certain sorts—complex images, expectations, etc. . . . Sentences signify something other than themselves, which can be the same when the sentences differ. That this something must be psychological (or physiological) is made evident by the fact that propositions can be false."[20] Thus it seems that Russell chooses a subjective, mental explicatum for the concept of proposition only or mainly for the reason that, in his opinion, there is no other way of overcoming the difficulty connected with false propositions.

[18] For the concepts of disjunctions or conjunctions of infinitely many propositions see [I], pp. 92 f.

[19] Russell, [Inquiry], p. 229 (chap. xiii, in sec. A). (Page numbers refer to the American edition; it seems that the British edition has, unfortunately, a different pagination.)

[20] *Ibid.*, pp. 237 f. (chap. xiii, end of sec. A).

I believe that it is possible to give an objective interpretation to the term 'proposition', which is still applicable in the case of false sentences. Any proposition must be regarded as a complex entity, consisting of component entities, which, in their turn, may be simple or again complex. Even if we assume that the ultimate components of a proposition must be exemplified, the whole complex, the proposition itself, need not be. The situation can perhaps best be made clear by its analogy with the situation concerning properties. As we have seen earlier (§ 4), a compound predicator, for example, 'H•T', may express an empty property, that is, one not exemplified by any individual. The components 'H' and 'T' express properties which are exemplified. The property expressed by the compound predicator is constituted out of the component properties in a logical structure indicated by the logical particles connecting the component predicators. Thus we see that the fact that some predicators are empty cannot prevent the explication of properties as objective entities. Analogously, the fact that some sentences are false does not exclude the explication of propositions as objective entities. Propositions, like complex properties, are complex entities; even if their ultimate components are exemplified, they themselves need not be. The difference between propositions and complex properties or other complex concepts is merely a difference in the logical type. Therefore, the kind of connection is different. In the case of our example 'H•T', the connection was that of conjunction. There are other logical connections which, applied to nonpropositional components of suitable types, result in propositions. Consider as an example the sentence 'Hs' of the system S_1; it consists of the predicator 'H' and the individual constant 's' combined by juxtaposition. Therefore, it expresses a complex intension of propositional type. Its two components are the intension of 'H', which is the property Human, and the intension of 's', which is, as we shall see later (§ 9), the individual concept Walter Scott. The logical connection of these two intensions is that of attribution or predication (expressed in S_1 simply by juxtaposition; its converse is expressed in certain other symbolic languages by 'ϵ' and in English by the copula 'is'). Thus the resultant intension of the sentence is the proposition that Scott is human. As an example of a different structure take '$(x)(Bx \supset Fx)$'. The intension of 'B' is the property Biped, that of 'F' is the property Featherless. These two properties are the components of the complex intension of the whole sentence. They are connected by the universal conditional connection, expressed, according to the rules of the system, by the way in which 'B' and 'F' are combined in the sentence with the help of three occurrences of a variable, two pairs of parentheses, and the conditional connective '⊃'. This kind of connection yields, if applied to two properties, a proposition. Thus the complex intension expressed by the sentence is the proposition that whatever is a biped is featherless. Each of the two component properties is exemplified by some individuals. Some of the sentences of the form '$\mathrm{B}x \supset \mathrm{F}x$' are exemplified by facts and hence true, but some of them are not. The whole intension is not exemplified; but it is, nevertheless, a proposition because it consists of exemplified components in a propositional structure; just as the intension of 'H•T', though empty, is a property because it consists of two exemplified components in the structure of a property. Thus F-false sentences, too, express propositions. Now we may go one step further. Consider the L-false sentence '(H•∼H)s'. It consists of the predicator 'H•∼H' and the individual constant 's', in the same combination as in the previous example 'Hs'. We have seen earlier (§ 4), that the predicator 'H•∼H', although L-empty, expresses a property, namely, the L-empty property Human And Non-Human. Therefore, the sentence mentioned expresses a complex intension resulting from combining this property with the individual concept Walter Scott by attribution. Thus this intension is the proposition that Scott is human and not human. Although this intension, like that of 'H•∼H', cannot possibly be exemplified, it still is a proposition. By going one step further in the analysis of this proposition we find as its components the property Human and the individual concept Walter Scott; these components are both exemplified, and they are combined in a structure of propositional type.

Generally speaking, it must perhaps be admitted that a designator can primarily express an intension only if it is exemplified. However, once we have some designators which have a primary intension, we can build compound designators out of them which express derivative, complex intensions, no matter whether these compound designators are exemplified or not. We do not need exemplifications in order to grasp their intensions, because the intension of a compound designator is determined, in virtue of the semantical rules of the system, by the intensions of the component designators and by the way in which these designators are combined.

It has been the purpose of the preceding remarks to facilitate the understanding of our conception of propositions. If, however, a reader should find these explanations more puzzling than clarifying, or even unacceptable, he may simply disregard them. They are not a necessary basis for the further discussions in this book; they will hardly be referred to again. It will be sufficient for nearly all our discussions involving propositions to assume that they are entities of any kind fulfilling the following two conditions: (1) to every sentence in a semantical system *S*, exactly one entity of this kind is assigned by the rules of *S;* (2) the same entity is assigned to two sentences in *S* if and only if these sentences are L-equivalent. If someone is in doubt as to whether there are any nonmental and extra-linguistic entities which fulfil these conditions, he may take as propositions certain linguistic entities which do so. We shall later see that, for example, certain classes of sentences in *S* may be taken (the L-equivalence classes, see remark at the end of § 33) or certain classes of classes of sentences in *S* (the ranges, see remark near the end of § 40).

§ 7. Individual Descriptions

An (individual) *description* is an expression of the form '$(\imath x)(\ldots x \ldots)$'; it means 'the one individual such that . . *x* . .'. If there is one and only one individual such that . . *x* . . , we say that the description satisfies the uniqueness condition. In this case the *descriptum*, i.e., the entity to which the description refers, is that one individual. Logicians differ in their interpretations of descriptions in cases where the uniqueness condition is not satisfied. The methods of Hilbert and Bernays and of Russell are here discussed; that of Frege will be discussed in the next section.

We use the term ***'individual'*** not for one particular kind of entity but, rather, relative to a language system *S*, for those entities which are taken as the elements of the universe of discourse in *S*, in other words, the entities of lowest level (we call it level zero) dealt with in *S*, no matter what these entities are. For one system the individuals may be physical things, for another space-time points, or numbers, or anything else. Consequently, we call the variables of level zero individual variables, the constants individual constants, and all expressions of this level, whether simple (variables and constants) or compound, ***individual expressions.*** The most important kinds of compound individual expressions are: (1) full expressions of functors (e.g., '$3 + 4$', where '$+$' is a functor and '3' and '4' are individual constants); within our systems, expressions of this kind occur only in S_3, not in S_1 and S_2; (2) individual descriptions. We shall use here the term 'description' mostly in the sense of 'individual description'. Descriptions of other types do not occur in our systems; a few remarks on them will be made at the end of § 8.

A ***description*** in S_1 has the form '$(\imath x)(\ldots x \ldots)$'; it is interpreted as 'the one individual *x* such that . . *x* . .'. '$(\imath x)$' is called an iota-operator; the scope '. . *x* . .' is a sentential matrix with '*x*' as a free variable. For example, '$(\imath x)(\mathrm{P}x \bullet \sim \mathrm{Q}x)$' means the same as 'the one individual which is P and not Q'.

The entity for which a description stands (if there is such an entity) will be called its ***descriptum;*** here, in the case of individual descriptions, the descriptum is an individual. With respect to a given description, there are two possible cases: either (1) there is exactly one individual which fulfils the condition expressed by the scope, or (2) this does not hold, that is, there are none or several such individuals. In the first case we shall say of the scope, and also of the whole description, that it satisfies the uniqueness condition:

7-1. *Definition.* Let '. . x . .' be a (sentential) matrix (in S_1) with 'x' as the only free variable. '. . x . .' (and '$(\imath x)$ (. . x . .)') satisfies the ***uniqueness condition*** (in S_1) $=_{Df}$ '$(\exists z)(x)[. . x . . \equiv (x \equiv z)]$' is true (in S_1). ('$x \equiv z$' means 'x is the same individual as z'; see 3-3.)

In the case of a description satisfying the uniqueness condition, there is general agreement among logicians with respect to its interpretation; the one individual satisfying the scope is taken as descriptum. In the other case, however, there is, so far, no agreement. Various methods have been proposed. We shall outline three of them, those proposed by Hilbert and Bernays (I), Russell (II), and Frege (III). Then we shall adopt Frege's method for our systems. It should be noticed that the various conceptions now to be discussed are not to be understood as different opinions, so that at least one of them must be wrong, but rather as different proposals. The different interpretations of descriptions are not meant as assertions about the meaning of phrases of the form 'the so-and-so' in English, but as proposals for an interpretation and, consequently, for deductive rules, concerning descriptions in symbolic systems. Therefore, there is no theoretical issue of right or wrong between the various conceptions, but only the practical question of the comparative convenience of different methods.

In order to make the following discussions more concrete, let us suppose that two (sentential) matrices are given, each with exactly one free variable; we indicate them here with the help of dots and dashes: '. . x . .' and '- - y - -' (e.g., 'Axw' and 'Hy'). We construct the description with the first as scope and substitute it for 'y' in the second:

7-2. '- - $(\imath x)$(. . x . .) - -'. (*Example:* 'H$(\imath x)$(Axw)'.)

Method I. ***Hilbert and Bernays,***[21] in a system with natural numbers as individuals, permit the use of a description only if it satisfies the uniqueness condition. Since the system is constructed as a calculus, not as a semantical system, the formula of uniqueness is required to be C-true (provable) instead of true. It seems that this method is quite con-

[21] [Grundlagen I], p. 384.

venient for practical work with a logico-arithmetical system; one uses a description only after he has proved the uniqueness. However, this method has a serious disadvantage, although of a chiefly theoretical nature: the rules of formation become indefinite, i.e., there is no general procedure for determining whether any given expression of the form 7-2 is a sentence of the system (no matter whether true or false, provable or not). For systems also containing factual sentences, the disadvantage would be still greater, because here the question of whether a given expression is a sentence or not would, in general, depend upon the contingency of facts.

Method II. ***Russell***[22] takes the whole expression 7-2 in any case as a sentence. The uniqueness condition is here taken not as a precondition for the sentential character of the expression but rather as one of the conditions for its truth—in other words, as part of its content. Thus the translation of 7-2 into M is as follows:

7-3. 'There is an individual y such that y is the only individual for which $..y..$ holds, and $--y--$' (for example, 'there is an individual y such that y is the only individual which is an author of Waverley, and y is human').

Hence, 7-2 is here interpreted as meaning the same as the following (with a certain restriction, see below):

7-4. '$(\exists y)[(x)(..x.. \equiv (x \equiv y)) \bullet --y--]$'. (In the example, '$(\exists y)[(x)(\mathrm{A}x\mathrm{w} \equiv (x \equiv y)) \bullet \mathrm{H}y]$'.)

In order to incorporate this interpretation into his system, Russell lays down a contextual definition for descriptions; 7-2 is the definiendum, 7-4 the definiens. If we prefer to take the iota-operator as primitive instead of defining it, we can reach the same result by framing the semantical rules in such a way that any two sentences of the forms 7-2 and 7-4 become L-equivalent.

In comparison with Hilbert's method, Russell's has the advantage that an expression of the form 7-2 is always a sentence. In comparison with Frege's method, which will soon be explained, it has the disadvantage that the rules for descriptions are not so simple as those for other individual expressions, especially those for individual constants. In particular, the inferences of specification, leading from '$(y)(--y--)$' to '$--\mathrm{a}--$', and of existential generalization, leading from '$--\mathrm{a}--$' to '$(\exists y)(--y--)$', are, in general, not valid if a description takes the place

[22] The reasons for this method are explained in detail by Russell in [Denoting]; it has been applied by Russell and Whitehead in the construction of the system of [P.M.], see I, 66 ff. and 173 ff.

of the individual constant 'a'; here the uniqueness sentence for the description must be taken as an additional premise. A further disadvantage of Russell's method is the following: A sentence like '$\sim Q(\iota x)(Px)$' can be transformed in two ways. Either this whole sentence is taken as 7-2 and transformed into the corresponding sentence of the form 7-4; or the part '$Q(\iota x)(Px)$' is taken as 7-2, transformed into the corresponding sentence of the form 7-4, and then prefixed again with the sign of negation. The two resulting sentences are not L-equivalent (in distinction to Frege's method); hence Russell has to lay down an additional convention, which determines for each case what is to be taken as the context 7-2.

§ 8. Frege's Method for Descriptions

We adopt for our systems a method proposed by Frege for interpreting individual descriptions in cases of nonuniqueness. This method consists in choosing once for all an individual to be taken as descriptum for all such cases.

Method III. ***Frege***[23] regards it as a defect in the logical structure of natural languages that in some cases an expression of the grammatical form 'the so-and-so' is a name[24] of one object while in other cases it is not; in our terminology: that some descriptions have a descriptum but others not. Therefore, he suggests that the rules of a language system should be constructed in such a way that every description has a descriptum. This requires certain conventions which are more or less arbitrary; but this disadvantage seems small in comparison with the gain in simplicity for the rules of the system. For instance, specification and existential generalization are here valid also for descriptions (at least in extensional contexts).

Frege's requirement can be fulfilled in various ways. The choice of a convenient procedure depends upon the particular features of the language system, especially upon the range of values of the variables in question. There are chiefly two methods which deserve consideration; we call them IIIa and IIIb. We shall explain them and then use IIIb for our systems.

Method IIIa. Frege[25] himself constructs a system without type difference between individuals and classes; that is to say, he counts both classes and their elements as objects, i.e., as values of the individual variables. To any of those descriptions which do not satisfy the condition of uniqueness he assigns as descriptum the class of those objects which fulfil

[23] [Sinn], pp. 39–42.

[24] For the question of English translations for Frege's terms, see below, p. 118, n. 21.

[25] [Grundgesetze], I, 19.

the scope. Thus different descriptions of this kind may have different descripta.

Method IIIb. A simpler procedure consists in selecting, once for all, a certain entity from the range of values of the variables in question and assigning it as descriptum to all descriptions which do not satisfy the condition of uniqueness. This has been done in various ways.

(i) If the individuals of the system are numbers, the number 0 seems to be the most natural choice. Frege[26] has already mentioned this possibility. It has been applied by Gödel[27] for his epsilon-operator and by myself[28] for the K-operator.

(ii) For variables to whose values the null class Λ belongs, this class seems to be the most convenient choice. Such a choice has been made by Quine,[29] in whose system there is, as in Frege's, no type difference between individuals and classes.

(iii) How can Method IIIb be applied to a language system whose individuals are physical things or events? At first glance, it seems impossible to make here an even moderately natural choice of an individual as common descriptum for all individual descriptions which do not satisfy the condition of uniqueness. To select, say, Napoleon would be just as arbitrary as to select this dust particle on my paper. However, a natural solution offers itself if we construct the system in such a way that the spatiotemporal part-whole relation is one of its concepts.[30] Every individual in such a system, that is, every thing or event, corresponds to a class of space-time points in a system with space-time points as individuals. Therefore, it is possible, although not customary in the ordinary language, to count among the things also the *null thing*, which corresponds to the null class of space-time points. In the language system of things it is characterized as that thing which is part of every thing.[31] Let us take 'a_0'

[26] [Sinn], p. 42 n.

[27] K. Gödel, "Ueber formal unentscheidbare Sätze der Principia Mathematica und verwandter Systeme", *Monatshefte f. Math. u. Physik*, XXXVIII (1931), 173–98.

[28] [Syntax], § 7.

[29] [M.L.], p. 147.

[30] This is, for instance, the case with the following systems: a system for certain biological concepts by J. H. Woodger (*The Axiomatic Method in Biology* [1937]; *The Technique of Theory Construction* ["International Encyclopedia of Unified Science", Vol. II, No. 5 (1939)]); a calculus of individuals by H. S. Leonard and N. Goodman ("The Calculus of Individuals and Its Uses", *Journal of Symbolic Logic*, V [1940], 45–55); and a general system of logic recently constructed by R. M. Martin ("A Homogeneous System for Formal Logic", *Journal of Symbolic Logic*, VIII [1943], 1–23), where the customary symbol of inclusion and the term 'inclusion' apparently refer to the part-whole relation among things.

[31] In the system by Martin mentioned in the preceding footnote the null thing is indeed introduced (see *op. cit.*, p. 3, and D7, p. 9), while in the paper by Leonard and Goodman there is an explicit "refusal to postulate a null element" (*op. cit.*, p. 46).

as the name for the null thing; the other things may be called non-null things. If a system S includes a_0 among its individuals, then a_0 seems a natural and convenient choice as descriptum for those descriptions which do not satisfy the uniqueness condition. It is true that this procedure requires certain deviations from the ordinary language for the forms of sentences in S; but these deviations are smaller than we might expect at first glance. For most of the universal and existential sentences, the translation into S is straightforward, i.e., without change in structure; in other cases 'non-null' must be inserted. [*Examples:* The sentence 'There is no thing which is identical with the king of France in 1905' is translated into a sentence of S of the form 'There is no non-null thing . . .'. On the other hand, no such change in form is necessary for the sentence 'All men are mortal' and not even for 'There is no man who is identical with the king of France in 1905', because it follows from any suitably framed definition for 'man' that every man is a non-null thing.]

In our further discussions we assume for our system S_1 that Frege's Method IIIb is applied and that the individual constant 'a*' is used for the common descriptum of all descriptions which do not satisfy the uniqueness condition. We leave it open which individual is meant by 'a*'; it may be the null thing a_0, if this belongs to the individuals in S_1; it may be 0, if numbers belong to the individuals (as, for instance, in S_3), but it may as well be any other individual. Consequently, a sentence containing a description is now interpreted in a way different from Russell's. The translation of 7-2 into M is now as follows (instead of 7-3):

8-1. 'Either there is an individual y such that y is the only individual for which . . y . . holds, and - - y - -; or there is no such individual, and - - a* - -'. [In the previous example: 'Either there is an individual y such that y is the only author of Waverley, and y is human; or there is no such individual y (that is to say, there is either no author or several authors of Waverley), and a* is human'.]

Hence, the sentence 7-2 containing the description is L-equivalent in S_1 to the following (instead of to 7-4):

8-2. '$(\exists y)[(x)(\,..\,x\,..\, \equiv (x \equiv y)) \bullet \text{- -}\,y\,\text{- -}] \vee [\sim(\exists y)(x)(\,..\,x\,..\, \equiv (x \equiv y)) \bullet \text{- -}\,a^*\,\text{- -}]$'. (In the example: '$(\exists y)[(x)(\mathrm{A}x\mathrm{w} \equiv (x \equiv y)) \bullet \mathrm{H}y] \vee [\sim(\exists y)(x)(\mathrm{A}x\mathrm{w} \equiv (x \equiv y)) \bullet \mathrm{Ha}^*]$'.)

Here again, as in the case of Russell's method, we may set up either a contextual definition for 7-2 with 8-2 as definiens, or semantical rules for the iota-operator as a primitive sign such that 7-2 becomes L-equivalent to 8-2.

The accompanying table gives a survey of the various methods just explained for dealing with descriptions in the case of nonuniqueness. The case of uniqueness is not represented because its treatment is the same with all authors.

INTERPRETATION OF DESCRIPTIONS IN THE CASE OF NONUNIQUENESS

HILBERT-BERNAYS	RUSSELL	FREGE (a)	FREGE (b)	QUINE	SYSTEM OF THINGS	SYSTEM S_1
Method I	Method II	Method IIIa	Method IIIb	Method IIIb	Method IIIb	Method IIIb
Description is meaningless	No descriptum; the sentence is meaningful but false	$\hat{x}(..x..)$	o	Λ	Null thing a_0	a*

Some brief remarks may be made on *descriptions with variables of other than individual type*, especially predicator variables, functor variables, and sentential variables. (This is a digression from the study of our systems S_1, etc., which contain only individual variables.) Here it is easy to make a natural choice of a value of the variable as a descriptum for those descriptions which do not satisfy the condition of uniqueness. If an individual has been chosen as a* (it may be a_0 or o or anything else), then we might call one entity in every type the null entity of that type, in the following way: In the type of individuals it would be a*; in any predicator type, the null class or null relation of that type, e.g., for level one and degree one the null class Λ; in the type of propositions, the L-false proposition; in any type of functions, that function which has as value for all arguments the null entity of the type in question. Then we may take as descriptum in the case of nonuniqueness the null entity of the type of the description variable.

For the sake of simplicity, the following explanations are restricted to extensional systems. Let '*f*' and '*g*' be predicator variables of level one and degree one. Let '$--(\imath f)(..f..)--$' indicate, in analogy to 7-2, a sentence containing a description of the type of '*f*', hence a description for a class or property. This sentence is L-equivalent to the following, in analogy to 8-2:

'$(\exists g)[(f)(..f.. \equiv (f \equiv g)) \bullet --g--] \vee [\sim(\exists g)(f)(..f.. \equiv (f \equiv g)) \bullet --\Lambda--]$'.

The uniqueness condition here occurring says that there is a property g, such that for those f and only those, which are equivalent to g, $..f..$; in other words, there is exactly one class g such that $..g...$ Hence here the uniqueness applies to extensions, not to intensions. This is in analogy to 7-1 and 7-3; for, as we shall see later, the extensions of individual expressions are individuals.

However, if the system contains lambda-operators for the formation of predicators, then descriptions with predicator variables are not necessary, they can be replaced by lambda-expressions. In this case we can transform not only a sentence containing the description as in the earlier case but the description itself into an L-equivalent expression. The description '$(\iota f)(..f..)$' is L-equivalent to the lambda-predicator '$(\lambda x)[(\exists g)((f)[..f.. \equiv (f \equiv g)] \bullet gx)]$'.

In a similar way, for every description of a function (containing an iota-operator with a functor variable) there is an L-equivalent functor formed with a lambda-operator. And for every description containing an iota-operator with a sentential variable there is an L-equivalent sentence without an iota-operator; however, in an extensional system these descriptions with sentential variables are rather useless anyway.

In view of these results, it seems convenient in the primitive notation of a system (at least in an extensional one) to use the iota-operator, if at all, for individual descriptions only, and then to use the lambda-operator for the formation of predicators and functors.[32]

§ 9. Extensions and Intensions of Individual Expressions

> It is found to be in accord with our earlier conventions, to take as the extension of an individual expression the individual to which it refers. The intension of an individual expression is a concept of a new kind; it is called an *individual concept*.

Let us consider some examples of F-equivalence and L-equivalence of individual expressions. We assume the following as a historical fact:

9-1. *Assumption.* There is one and only one individual which is an author of Waverley, and this individual is the same as Walter Scott.

Then the descriptum of '$(\iota x)(\text{A}x\text{w})$' is that individual which is author of Waverley and not a* and '$(\iota x)(\text{A}x\text{w}) \equiv \text{s}$' is, according to the rule 3-3, true, but not L-true; hence it is F-true. This leads to the following result, according to the definitions 3-5:

[32] Several forms of systems with predicators and functors built with lambda-operators have been constructed by Church, see especially *The Calculi of Lambda-Conversion* ("Ann. of Math. Studies", No. 6 [1941]).

9-2. '$(\imath x)(\mathrm{A}x\mathrm{w})$' is equivalent to 's', but not L-equivalent, hence F-equivalent.

On the other hand, let us compare the two descriptions '$(\imath x)(\mathrm{H}x \bullet \mathrm{A}x\mathrm{w})$' and '$(\imath x)(\mathrm{RA}x \bullet \mathrm{A}x\mathrm{w})$'. Let us see what we can find out about them if we make use of the rules of S_1, especially 1-2, but not of any historical or other factual knowledge. If there is exactly one individual which is both human —or, which means the same, a rational animal—and an author of Waverley, then the descriptum of each of the two descriptions is this individual; otherwise the descriptum of each is a*. Thus, in either case, the descriptum of the first description is the same individual as that of the second. Hence, according to rule 3-3, the sentence '$(\imath x)(\mathrm{H}x \bullet \mathrm{A}x\mathrm{w}) \equiv (\imath x)(\mathrm{RA}x \bullet \mathrm{A}x\mathrm{w})$' is true; it is, moreover, L-true because we have shown its truth by using merely the semantical rules. Therefore, the two descriptions are L-equivalent.

We found earlier that individual expressions are equivalent if and only if they are expressions for the same individual (3-12). Hence, according to the definition of identity of extensions (5-1), individual expressions have the same extension if and only if they are expressions for the same individual. Therefore, it seems natural to regard as extensions of individual expressions the individuals themselves:

9-3. The *extension of an individual expression* is the individual to which it refers (hence the descriptum, if it is a description).

Since we adopted Frege's method, every description has exactly one descriptum. Hence, on the basis of the convention just made, there is no ambiguity with respect to the extension of an individual expression. For instance, the extension of 's' is the individual Walter Scott, and the same holds for each of the three descriptions discussed above as examples. If there were none or several authors of Waverley, then the extension of '$(\imath x)(\mathrm{A}x\mathrm{w})$' would be the individual a*.

Now let us look for entities which we might regard as intensions of individual expressions. According to our definition for the identity of intensions (5-2), the intension must be something that L-equivalent individual expressions (for example, the two descriptions above containing 'H' and 'RA') have in common. We have earlier found entities which seemed suitable as intensions of designators of other types; for sentences, propositions; for predicators, properties or relations; for functors, functions. Thus, in these cases, the intensions are those entities which are sometimes regarded as the meanings of the expressions in question; and, in the case of predicators and functors, the intensions are concepts of certain types. Now it seems to me a natural procedure, in the case of individual expressions, likewise to speak of concepts, but of concepts of a particular type, namely, the individual type. Although it is not altogether customary to speak here of concepts in this sense, still it does not seem to deviate too much from ordinary usage. I propose to use the term ***'individual concept'*** for this type of concept. Thus we say:

9-4. The *intension of an individual expression* is the individual concept expressed by it.

Examples:

9-5. The intension of 's' is the individual concept Walter Scott.

9-6. The intension of '$(\imath x)(\mathrm{A}x\mathrm{w})$' is the individual concept The Author Of Waverley.

(Here, and further on, in translating descriptions into M, we omit for brevity the phrase 'or a*, if there is not exactly one such individual'.) Instead of saying in the customary but ambiguous terminology that the two L-equivalent descriptions discussed above have the same meaning, we say now that they have the same intension and that their common intension is the individual concept The Human Author Of Waverley, which is the same as the individual concept The Rational Animal Author Of Waverley. On the other hand, the following are three different individual concepts: the one just mentioned, the individual concept Walter Scott, and the individual concept The Author Of Waverley. Here again the intensions of given expressions, and the identity or nonidentity of these intensions, can be determined on the basis of the semantical rules alone.

We have seen earlier how a sentence containing a predicator can be translated into M, that is, English, in different ways. Thus, for the sentence 'Hs', we had, in addition to the simple translation 'Scott is human', two more explicit translations, one of which used the term 'property' and the other the term 'class' (see 4-2 and 4-3). In these two explicit translations, 's' was still simply translated by 'Scott'. Now, however, we have seen that, corresponding to the distinction between classes and properties, we have in the case of individual expressions the distinction between individuals and individual concepts. Hence, we may use in M instead of 'Scott' the more explicit phrases 'the individual Scott' and 'the individual concept Scott'. Since the distinction is perhaps clearer for a description than for an individual constant, let us take, instead of 'Hs', the sentence '$\mathrm{H}(\imath x)(\mathrm{A}x\mathrm{w})$'. In addition to the simple translation 'the author of Waverley is human', we have here four more explicit translations in which both to 'The Author Of Waverley' and to 'Human' a characterizing word is added. Two of these translations are pure, two mixed. Of the two pure translations, the first contains two references to extensions, and the second two references to intensions; these translations are as follows:

'The individual The Author Of Waverley belongs to the class Human'.

'The individual concept The Author Of Waverley is subsumable under the property Human'.

Since it is not customary to speak about individual concepts, there is no word in customary usage for the relation between an individual concept and a property corresponding to the element-relation between an individual and a class; we have used here for this relation the word 'subsumable' (in the sense of 'truly subsumable'), but we shall not use it further on. Of the two mixed translations, which contain a reference to an extension and a reference to an intension, we shall give at least one, because it is not too far from customary usage:

'The individual The Author Of Waverley has the property Human'.

Thus we find here a multiplicity of possible translations into M, some of them rather cumbersome and strange-looking. This multiplicity seems inevitable as long as we wish to distinguish explicitly between classes and properties and between individuals and individual expressions. The problem of whether and by which means this apparent multiplicity of entities and the corresponding multiplicity of formulations can be reduced will be discussed later (§§ 33 f.).

§ 10. Variables

We found earlier that the extension of a predicator 'P' is a class, and its intension is a property. Therefore, a variable of the same type (e.g., '*f*') refers both to classes and to properties; we say that classes are its *value extensions*, and properties its *value intensions*. Analogously, for a variable of the type of sentences (e.g., '*p*'), the value extensions are truth-values, and the value intensions are propositions. Finally, the value extensions of an individual variable (e.g., '*x*') are individuals, and its value intensions are individual concepts.

Quine has repeatedly pointed out the important fact that, if we wish to find out what kind of entities somebody recognizes, we have to look more at the variables he uses than at the constants and closed expressions. "The ontology to which one's use of language commits him comprises simply the objects that he treats as falling . . . within the range of values of his variables."[33] I am essentially in agreement with this view, as I shall presently explain. But, first, I wish to indicate a doubt concerning Quine's *formulation;* I am not quite clear whether the point raised is not perhaps of a

[33] [Notes], p. 118; see also his [Designation].

merely terminological nature. I should prefer not to use the word '*ontology*' for the recognition of entities by the admission of variables. This use seems to me to be at least misleading; it might be understood as implying that the decision to use certain kinds of variables must be based on ontological, metaphysical convictions. In my view, however, the choice of a certain language structure and, in particular, the decision to use certain types of variables is a practical decision like the choice of an instrument; it depends chiefly upon the purposes for which the instrument—here the language—is intended to be used and upon the properties of the instrument. I admit that the choice of a language suitable for the purposes of physics and mathematics involves problems quite different from those involved in the choice of a suitable motor for a freight airplane; but, in a sense, both are engineering problems, and I fail to see why metaphysics should enter into the first any more than into the second. Furthermore, I, like many other empiricists, regard the alleged questions and answers occurring in the traditional realism-nominalism controversy, concerning the ontological reality of universals or any other kind of entities, as pseudo-questions and pseudo-statements devoid of cognitive meaning. I agree, of course, with Quine that the problem of "Nominalism" as he interprets it[34] is a meaningful problem; it is the question of whether all natural science can be expressed in a "nominalistic" language, that is, one containing only individual variables whose values are concrete objects, not classes, properties, and the like. However, I am doubtful whether it is advisable to transfer to this new problem in logic or semantics the label 'nominalism' which stems from an old metaphysical problem.

The sense in which I agree with Quine's thesis that "to be is to be the value of a variable" will become clear by the following example: Suppose somebody constructs a language not only as a subject matter of theoretical investigations but for the purpose of communication. Suppose, further, that he decides to use in this language variables 'm', 'n', etc., for which all (natural) numerical expressions (e.g., '0', '3', '2 + 3', etc.) and only those are substitutable. We see from this decision that he recognizes natural numbers in this sense: he is willing to speak not only about particular numbers (e.g., '7 is a prime number') but also—and this is the decisive point—about numbers in general. He will, for example, make statements like: 'for every m and n, $m + n = n + m$' and 'there is an m between 7 and 13 which is prime'. The latter sentence speaks of the existence of a prime number. However, the concept of existence here has nothing to do with the ontological concept of existence or reality. The sen-

[34] [Designation], p. 708.

tence mentioned means just the same as 'it is not the case that for every *m* between 7 and 13, *m* is not prime'. By the same token, we see, furthermore, that the user of the language is willing to recognize the concept Number. Generally speaking, if a language (of ordinary structure) contains certain variables, then we can define in it a designator for the range of values of those variables. In the present case, the definition is: " 'Number' for '$(\lambda m)(m = m)$' " or, if the language in question does not contain abstraction operators, " 'Number(m)' for '$m = m$' ". [In the definiens, any matrix '. . *m* . .' may be used which is L-universal, that is, such that '(m)(. . *m* . .)' is L-true.] It is important to emphasize the point just made that, once you admit certain variables, you are bound to admit the corresponding universal concept. It seems to me that some philosophers (not Quine) overlook this fact; they do not hesitate to admit into the language of science variables of the customary kinds, like sentential variables ('*p*', '*q*', etc.), numerical variables, perhaps also predicator variables at least of level one, and other kinds; at the same time, however, they feel strong misgivings against words like 'proposition', 'number', 'property' (or 'class'), 'function', etc., because they suspect in these words the danger of an absolutist metaphysics. In my view, however, the accusation of an absolutist metaphysics or of illegitimate hypostatizations with respect to a certain kind of entities, say propositions, cannot be made against an author, merely on the basis of the fact that he uses variables of the type in question (e.g., '*p*', etc.) and the corresponding universal word ('proposition'); it must be based, instead, on an analysis of the statements or pseudo-statements which he makes with the help of those signs.

Quine's thesis and my remarks in connection with it concern the language which somebody not only analyzes but uses, hence, with respect to semantical discussions, the metalanguage. Now let us look at the role of variables in an object language *S*. If *S* is given, then a metalanguage *M* intended for the semantical analysis of *S* must be rich enough in relation to *S*. In particular, *M* must contain variables whose ranges of values cover those of all variables in *S* (and, as Tarski has shown, even go beyond this in order to make possible the definition of 'true in *S*'). Let us further presuppose here, as in the previous discussions, that *M* enables us to speak in general terms about the extensions and intensions of predicators, sentences, and individual expressions of *S*.

Let *S* (in distinction to S_1) contain not only individual variables but also those of other types. Let us begin with variables '*f*', '*g*', etc., of the type of predicators of level one and degree one. With respect to a predicator, say 'H' in S_1, we have distinguished between its extension, the class

Human, and its intension, the property Human. A sentence '. . H . .' containing 'H' can be translated into *M* in different ways; we may use either the word 'human' alone or the phrase 'the class Human' or 'the property Human' (see, as an example, the translations of 'Hs' in § 4); we have seen that this involves merely a difference in formulation. Now in *S*, we can deduce from '. . H . .' the existential sentence '$(\exists f)(..f..)$'. For the translation of this sentence into *M* we have again three forms, corresponding to the three forms mentioned for the transla tion of '. . H . .':

(i) 'There is an *f* such that . . *f* . .',
(ii) 'There is a class *f* such that . . *f* . .',
(iii) 'There is a property *f* such that . . *f* . .'.

As 'H' is an expression both for the class Human and for the property Human, '*f*' is thus a variable both for classes and for properties. Since we regarded the class Human as the extension of 'H', we shall now regard it as one of the ***value extensions*** of '*f*'; and, analogously, we take the property Human as one of the ***value intensions*** of '*f*'. Let us call the closed expressions substitutable for a certain variable of any kind the ***value expressions*** of that variable. Then the following holds generally, for variables of any kind.

10-1. The extension of a value expression of a variable is one of the value extensions of that variable.

10-2. The intension of a value expression of a variable is one of the value intensions of that variable.

For variables of the type of sentences, say '*p*', '*q*', etc., the situation is analogous. Their value extensions are truth-values; their value intensions, propositions. Let '. . Hs . .' be a sentence containing 'Hs' as a proper subsentence. We may translate '. . Hs . .' into *M* in various ways. One possible translation contains simply the phrase '(that) Scott is human'. Of the two more explicit translations, one contains the phrase 'the truth-value that Scott is human', and the other 'the proposition that Scott is human', in accord with our earlier results concerning the extension and the intension of 'Hs' (6-3 and 6-4). Now in *S*, we may infer from '. . Hs . .' the existential sentence '$(\exists p)(..p..)$'. Corresponding to the three translations of '. . Hs . .', we have three translations of this existential sentence:

(i) 'There is a *p* such that . . *p* . .',
(ii) 'There is a truth-value *p* such that . . *p* . .',
(iii) 'There is a proposition *p* such that . . *p* . .'.

The treatment of individual variables is not essentially different from that of the other kinds of variables. But, owing to the unfamiliarity of individual concepts, our conception here may seem less natural at first glance. We considered earlier the sentence 'H($\imath x$)(Axw)' containing a description. In addition to the simple translation 'the author of Waverley is human', we had several more explicit translations containing the phrases 'the individual' and 'the individual concept' (at the end of the preceding section). From the sentence with the description (or from the simpler sentence 'Hs') we may deduce '($\exists x$)(Hx)'. Corresponding to the earlier translations of the former sentence, we have the following translations of this existential sentence:

(i) 'There is an x such that x is human'.
(ii) 'There is an individual x such that x belongs to the class Human'.
(iii) 'There is an individual concept x such that x is subsumable under the property Human'.
(iv) 'There is an individual x such that x has the property Human'.

Thus the value extensions of individual variables are individuals, their value intensions are individual concepts. The multiplicity of the formulations and the strangeness of some of them are the same here as in the preceding section. Our later attempt at a simplification will apply to the present situation, too.

§ 11. Extensional and Intensional Contexts

An expression occurring within a sentence is said to be *interchangeable* with another expression if the truth-value of the sentence remains unchanged when the first expression is replaced by the second. If, moreover, the intension of the sentence remains unchanged, the two expressions are said to be *L-interchangeable*. We say that a sentence is *extensional* with respect to an expression occurring in it or that the expression occurs in the sentence *within an extensional context*, if the expression is interchangeable at this place with every other expression equivalent to it. We say that the sentence is *intensional* with respect to the expression, or that the expression occurs *within an intensional context*, if the context is not extensional and the expression is L-interchangeable at this place with every other expression L-equivalent to it. (The definitions actually given in this section are wider than here indicated; they refer not only to sentences but to designators of any type.) It is found, in accordance with customary conceptions, that all sentences of the system S_1, which contains only the ordinary connectives and quantifiers but no modal signs, are extensional and that a sentence in S_2 of the form 'N(. . .)', where 'N' is a sign for logical necessity, is intensional.

Suppose that we replace an expression (designator or not) which occurs within a designator by another expression. It may happen that the extension of the designator is not thereby changed; in this case we call the two expressions interchangeable within the designator. If, moreover, the intension of the designator remains unchanged, we say that the two expressions are L-interchangeable within the designator. The subsequent definitions for these concepts in technical terms (11-1a) refer not to extension and intension but, instead, to equivalence and L-equivalence. Two further concepts are defined (11-1b), which apply to the case in which the conditions mentioned are fulfilled for all sentences. The system S to which these and the later definitions (11-2) refer may be one of our systems S_1, S_2, S_3, or a similar system with the same types of designators; it is supposed that S contains descriptive predicates, and hence factual sentences,[35] and also individual descriptions with those predicates. [Thus S may be PM′, but not PM, in § 26; it may be ML′, but not ML, in § 25.] S may, in distinction to our systems, also contain variables for the non-individual types of designators.

11-1. *Definitions*

a. An occurrence of the expression $\mathfrak{A}_j$ *within the expression* $\mathfrak{A}_i$ is (1) ***interchangeable,*** (2) ***L-interchangeable*** with $\mathfrak{A}_j'$ (in S) $=_{\mathrm{Df}}$ $\mathfrak{A}_i$ is a designator and is (1) equivalent, (2) L-equivalent to the expression $\mathfrak{A}_i'$ constructed out of $\mathfrak{A}_i$ by replacing the occurrence of $\mathfrak{A}_j$ in question by $\mathfrak{A}_j'$.

b. $\mathfrak{A}_j$ is (1) ***interchangeable,*** (2) ***L-interchangeable*** with $\mathfrak{A}_j'$ *in the system* S $=_{\mathrm{Df}}$ any occurrence of $\mathfrak{A}_j$ within any sentence of S is (1) interchangeable, (2) L-interchangeable with $\mathfrak{A}_j'$.

Consider a particular occurrence of a designator $\mathfrak{A}_j$ within a designator $\mathfrak{A}_i$. The situation may be such that the extension of $\mathfrak{A}_i$ depends merely upon the extension of $\mathfrak{A}_j$, that is to say, it remains unchanged if $\mathfrak{A}_j$ is replaced by any other expression with the same extension. In this case we shall say that $\mathfrak{A}_i$ is *extensional* with respect to that occurrence of $\mathfrak{A}_j$ (11-2a). We must here refer to a particular occurrence; for, if $\mathfrak{A}_i$ contains

[35] The fact that a restriction of this kind is necessary was pointed out to me by Alonzo Church. If S is a system of modal logic which, like Lewis' system of strict implication, contains no descriptive predicates and hence no factual sentences, then any two equivalent sentences are L-equivalent and hence are L-interchangeable even within a modal sentence of the form 'N(. . .)'. Thus the latter sentence would fulfil the condition of extensionality as stated below in 11-2b; in fact, however, a modal sentence is, of course, to be regarded as intensional in the customary sense. To state definitions of 'extensional' and 'intensional' which are applicable also to systems containing only L-determinate sentences or no closed sentences at all, it would be necessary to refer not only to closed designators occurring as parts but also to the values of the designator variables and to the corresponding values of propositional functions expressed by matrices (for example, to the values of 'p' and the corresponding values of 'Np'). In order to avoid this complication in our present discussion, we restrict the systems S as indicated in the text.

several occurrences of $\mathfrak{A}_j$, it may happen that one occurrence fulfils the above condition, while another does not. If the condition is fulfilled, we shall also say sometimes that $\mathfrak{A}_j$ occurs within $\mathfrak{A}_i$ at the place in question in an *extensional context*.

11-2. *Definitions*

a. The *expression* $\mathfrak{A}_i$ is ***extensional*** *with respect to a certain occurrence* of $\mathfrak{A}_j$ within $\mathfrak{A}_i$ (in the system S) $=_{\text{Df}}$ $\mathfrak{A}_i$ and $\mathfrak{A}_j$ are designators; the occurrence in question of $\mathfrak{A}_j$ within $\mathfrak{A}_i$ is interchangeable with any expression equivalent to $\mathfrak{A}_j$ (in S).

b. The *expression* $\mathfrak{A}_i$ is ***extensional*** (in S) $=_{\text{Df}}$ $\mathfrak{A}_i$ is a designator (in S); $\mathfrak{A}_i$ is extensional with respect to any occurrence of a designator within $\mathfrak{A}_i$ (in S).

c. The *semantical system* S is ***extensional*** $=_{\text{Df}}$ every sentence in S is extensional.

If the condition in 11-2a or b or c is not fulfilled, we shall use the term '*nonextensional*'. The term '*intensional*' (11-3) will be used not, as is sometimes done, as synonymous with 'nonextensional', but in a narrower sense, namely, in those cases in which the condition of extensionality is not fulfilled but the analogous condition with respect to intension is fulfilled. The latter condition means that the intension of the whole remains unchanged if the subexpression is replaced by one with the same intension; the technical definition (11-3) does not refer to intension but uses, instead, the concepts of L-equivalence and L-interchangeability.

11-3. *Definitions*

a. The *expression* $\mathfrak{A}_i$ is ***intensional*** *with respect to a certain occurrence* of $\mathfrak{A}_j$ within $\mathfrak{A}_i$ (in S) $=_{\text{Df}}$ $\mathfrak{A}_i$ and $\mathfrak{A}_j$ are designators; $\mathfrak{A}_i$ is not extensional with respect to the occurrence in question of $\mathfrak{A}_j$ within $\mathfrak{A}_i$; this occurrence of $\mathfrak{A}_j$ within $\mathfrak{A}_i$ is L-interchangeable with any expression L-equivalent to $\mathfrak{A}_j$ (in S).

b. The *expression* $\mathfrak{A}_i$ is ***intensional*** (in S) $=_{\text{Df}}$ $\mathfrak{A}_i$ is a designator; $\mathfrak{A}_i$ is, with respect to any occurrence of a designator within $\mathfrak{A}_i$, either extensional or intensional, and is intensional with respect to at least one occurrence of a designator.

c. The *semantical system* S is ***intensional*** $=_{\text{Df}}$ every sentence in S is either extensional or intensional, and at least one is intensional.

We shall sometimes call a sentential connective or a predicator constant extensional, if every full sentence of it is extensional with respect to the argument expressions; and we shall use the term 'intensional' analogously.

Note that the terms 'extension' and 'intension' occur only in the informal explanations and not in the definitions 11-1, 11-2, and 11-3 themselves. Thus these definitions do not presuppose any problematic entities. They use, instead, the terms 'equivalent' and 'L-equivalent', which, as mentioned earlier (§ 5), are unproblematic and can be defined in an exact way.

The terms 'interchangeable', 'L-interchangeable', 'extensional', and 'intensional' have been defined here in a general way so that the whole expression $\mathfrak{A}_i$ may be a designator of any of the types occurring in our systems. These terms find their most important application, however, in those cases in which $\mathfrak{A}_i$ is a sentence; and in our further discussions we shall use them chiefly for cases of this kind.

The concepts just defined will become clearer with some examples. The whole expression $\mathfrak{A}_i$ is a sentence in all these examples. The subexpression, $\mathfrak{A}_j$, is, in the first three examples, a sentence; in the later examples a designator of another type.

Example I. A sentence '. . V - -' is *extensional* with respect to either of its components. And, generally, as is well known, any full sentence of the ordinary connectives, '$\sim$', 'V', '$\bullet$', '$\supset$', and '$\equiv$', is extensional with respect to its (immediate) component or components. These connectives and the connections for which they stand are, indeed, often called extensional;[36] following Russell, the connections are usually called truth-functions.

Example II. Anticipating later explanations (chap. v), let us use here the system S_2, which contains the signs of S_1 and, in addition, 'N' as a modal sign for logical necessity in such a way that, if '. . .' is any L-true sentence, 'N(. . .)' is true and, moreover, L-true; and if '. . .' is any sentence not L-true, then 'N(. . .)' is false and moreover L-false (see 39-3). Let 'C' be an abbreviation for an F-true sentence (e.g., for 'Hs'); then 'C' is true but not L-true. As is well known (see the example following 2-2) 'C V$\sim$ C' is L-true. Hence:

11-4. 'C' and 'C V$\sim$C' are equivalent but not L-equivalent.

According to the given explanations for 'N', we have:

11-5. 'N(C V$\sim$C)' is true and, moreover, L-true.

On the other hand, since 'C' is not L-true, 'N(C)' is false. Therefore, 'N(C V$\sim$C)' and 'N(C)' are not equivalent. It follows, according to the definition 11-1a, that the occurrence of 'C' within 'N(C)' is not inter-

[36] The concept of extensionality of connections and connectives and the corresponding concept of L-extensionality are discussed in more detail in [II], § 12.

changeable with 'C ∨ ~ C'. This, together with 11-4 and the definition 11-2a, leads to the following result:

11-6. 'N(C)' is *nonextensional* with respect to 'C'.

This result is well known; generally, full sentences of modal signs are nonextensional with respect to their components; in customary terms, modalities are not truth-functions.[37] The same consideration shows that the occurrence of 'C ∨ ~ C' within 'N(C ∨ ~ C)' is not interchangeable with 'C'. Thus we obtain (again with 11-4):

11-7. 'N(C ∨ ~ C)' is *nonextensional* with respect to the subsentence 'C ∨ ~ C'.

Further, let 'D' be any sentence L-equivalent to 'C ∨ ~ C'. Then 'D' is likewise L-true; and hence 'N(D)', too. We found that 'N(C ∨ ~ C)' is L-true (11-5). Since any two L-true sentences hold in the same state-descriptions (2-2), they are L-equivalent to each other (2-6). Thus 'N(C ∨ ~ C)' and 'N(D)' are L-equivalent. Therefore, according to the definition 11-1a, the occurrence of 'C ∨ ~ C' within 'N(C ∨ ~ C)' is L-interchangeable with any sentence which is L-equivalent to 'C ∨ ~ C'. This, together with 11-7 and the definition 11-3a, yields:

11-8. 'N(C ∨ ~ C)' is *intensional* with respect to the subsentence 'C ∨ ~ C'.

Example III. The sentence 'Hs' is true in S_1; it remains true if 'H' is replaced by any equivalent predicator, for instance, by 'F•B'; and likewise if 's' is replaced by any equivalent individual expression, for instance, by the description '$(\imath x)$(Axw)' (9-2). Therefore:

[37] The results 11-6 and 11-7 refute Church's opinion that (on a certain assumption, see below) "Carnap's definition of 'extensional' fails in that under it every language (every semantical system) is extensional, even those which contain names of propositions and modal operators" ([Review C.], p. 304). The definition of 'extensional' here referred to is [I], D10-20 and D10-21, p. 43; it is essentially the same as 11-1 and 11-2 in the present section; however, the restriction to systems with factual sentences was omitted. Church is right in criticizing this omission (see n. 35). However, if the definition is applied to systems also containing factual sentences, like the example systems in my earlier book [I] and in the present book, then the definition seems to me to be adequate; at any rate, the examples here mentioned show that it is certainly not the case that under this definition (either in the earlier or in the present formulation) all sentences and all semantical systems fulfill the defining condition for extensionality. Church qualifies his statement by the following condition: "if the designatum of a sentence is always a truth-value." [Here the term 'designatum', as Church's preceding explanations show, is meant in the sense in which I shall use the term 'nominatum' in this book (§ 24); this sense is different from that in which I have used the term 'designatum' in [I], see below, § 37]. However, this qualification does not change the situation. Any assumption as to what are the designata (nominata) of sentences is irrelevant to the question of whether the examples stated in 11-6, 11-7, and 13-4 are extensional or not on the basis of my definition, because in this definition the concept of the designatum (nominatum) of a sentence is not used.

11-9. 'Hs' is extensional with respect to both 'H' and 's'.

Example IV. It can easily be shown that every sentence in S_1 constructed out of predicator constants (like those mentioned in rule 1-2), individual constants (like those mentioned in 1-1), connectives of the kind mentioned in Example I, universal and existential quantifiers, and iota- and lambda-operators is extensional with respect to any designators contained in it and hence is extensional (11-2b). S_1 is intended to contain only sentences constructed in this way. Therefore, according to definition 11-2c:

11-10. S_1 is an *extensional system.*

§ 12. The Principles of Interchangeability

Some theorems are stated concerning interchangeability and L-interchangeability in extensional and intensional contexts.

The following theorems, which we call principles of interchangeability, follow from our previous definitions of interchangeability and L-interchangeability (11-1), extensionality (11-2), and intensionality (11-3). The system *S*, to which the theorems of this section refer, is supposed to be either one of our systems S_1, S_2, S_3, or a similar system as specified earlier (see the explanation preceding 11-1).

12-1. ***First Principle of Interchangeability.*** Let $..\mathfrak{A}_j..$ be a sentence (in the system *S*) which is extensional with respect to a certain occurrence of the designator $\mathfrak{A}_j$, and $..\mathfrak{A}_k..$ the corresponding sentence with an occurrence of $\mathfrak{A}_k$ instead of that of $\mathfrak{A}_j$; analogously for '$..u..$' and '$..v..$' in c.

- **a.** If $\mathfrak{A}_j$ and $\mathfrak{A}_k$ are equivalent (in *S*), then the occurrence in question of $\mathfrak{A}_j$ within $..\mathfrak{A}_j..$ is interchangeable with $\mathfrak{A}_k$ (in *S*).
- **b.** $(\mathfrak{A}_j \equiv \mathfrak{A}_k) \supset (..\mathfrak{A}_j.. \equiv ..\mathfrak{A}_k..)$ is true (in *S*).
- **c.** Suppose that *S* contains variables for which $\mathfrak{A}_j$ and $\mathfrak{A}_k$ are substitutable, say 'u' and 'v'; then '$(u)(v)[(u \equiv v) \supset (..u.. \equiv ..v..)]$' is true (in *S*).

Statement 12-1a follows immediately from the definition 11-2a; and b and c follow from a by the general definition of equivalence (3-5a). The forms b and c have the advantage that here the principle is represented by a sentence in the object language *S* itself. The form c requires suitable variables. In the system S_1, for instance, form c is applicable only with individual variables and hence states only the interchangeability of individual expressions, while forms a and b apply also to predicators and sentences in S_1.

12-2. ***Second Principle of Interchangeability.*** Let . . $\mathfrak{A}_j$ . . be a sentence (in *S*) which is either extensional or intensional with respect to a certain occurrence of the designator $\mathfrak{A}_j$, and . . $\mathfrak{A}_k$ . . the corresponding sentence with $\mathfrak{A}_k$.

a. If $\mathfrak{A}_j$ and $\mathfrak{A}_k$ are L-equivalent (in *S*), then the occurrence in question of $\mathfrak{A}_j$ within . . $\mathfrak{A}_j$ . . is L-interchangeable and hence interchangeable with $\mathfrak{A}_k$ (in *S*).

Formulations b and c of the second principle analogous to 12-1b and c are possible only with the help of a modal sign, hence only with respect to a nonextensional language system. They will be given later (39-7b and c).

The following theorems follow from the two principles just stated, with the help of the definitions of extensional and intensional systems (11-2c and 11-3c):

12-3. Let *S* be an *extensional system* (for instance, S_1, see Example IV in § 11).

a. Equivalent expressions are interchangeable in *S*.

b. L-equivalent expressions are L-interchangeable in *S*.

Examples. **a.** Equivalence and therefore interchangeability in S_1 hold for the following pairs of expressions: (i) 'H' and 'F•B' (see 3-8); (ii) 'Hs' and '(F•B)(s)'; (iii) 's' and '(ɿ*x*)(A*x*w)' (see 9-2). **b.** L-equivalence and therefore L-interchangeability in S_1 hold for the following pairs of expressions: (i) 'H' and 'RA' (see 3-11); (ii) 'Hs' and 'RAs'; (iii) '(ɿ*x*)(H*x*•A*x*w)' and '(ɿ*x*)(RA*x*•A*x*w)' (see § 9).

12-4. Let *S* be an *intensional system* (for instance, S_2 with the modal sign 'N', see Example II in § 11 and § 39).

a. Equivalent expressions are interchangeable in *S*, except where they occur in an intensional context (for example, in the system S_2: except in a context of the form 'N(. . .)').

b. L-equivalent expressions are L-interchangeable in *S*.

Examples for S_2. **a.** Let 'C' be F-true, as in Example II, § 11. Then 'C' and 'C V ~ C' are equivalent (see 11-4). The sentence '(C V ~ C)• N(C V ~ C)' is true (see 11-5). Within this sentence the first occurrence of 'C V ~ C' is interchangeable with 'C', while the second is not. **b.** For the pairs of L-equivalent expressions in S_1 mentioned above, L-equivalence in S_2 and therefore L-interchangeability in S_2 likewise hold.

§ 13. Sentences about Beliefs

We study sentences of the form 'John believes that . . .'. If here the subsentence '. . .' is replaced by another sentence L-equivalent to it, then it may be that the whole sentence changes its truth-value. Therefore, the whole belief-sentence is neither extensional nor intensional with respect to the subsentence '. . .'. Consequently, an interpretation of belief-sentences as referring either to sentences or to propositions is not quite satisfactory. For a more adequate interpretation we need a relation between sentences which is still stronger than L-equivalence. Such a relation will be defined in the next section.

We found that ' . . . V - - -' is extensional with respect to the subsentence indicated by dots, and that 'N(. . .)' is intensional. Can there be a context which is neither extensional nor intensional? This would be the case if (but not only if) the replacement of a subsentence by an L-equivalent one changed the truth-value and hence also the intension of the whole sentence. In our systems this cannot occur; every sentence in S_1 (and likewise in S_3, to be explained later) is extensional, and every sentence in S_2 is either extensional or intensional. However, it is the case for a very important kind of sentence with psychological terms, like 'I believe that it will rain'. Although sentences of this kind seem to be quite clear and unproblematic at first glance and are, indeed, used and understood in everyday life without any difficulty, they have proved very puzzling to logicians who have tried to analyze them. Let us see whether we can throw some light upon them with the help of our semantical concepts.

In order to formulate examples, we take here, as our object language *S*, not a symbolic system but a part of the English language. We assume that *S* is similar in structure to S_1 except for containing the predicator '. . believes that - -' and some mathematical terms. We do not specify here the rules of *S;* we assume that the semantical rules of *S* are such that the predicator mentioned has its ordinary meaning; and, further, that our semantical concepts, especially 'true', 'L-true', 'equivalent', and 'L-equivalent', are defined for *S* in accord with our earlier conventions. Now we consider the following two belief-sentences; 'D' and 'D′' are here written as abbreviations for two sentences in *S* to be explained presently:

(i) 'John believes that D'.
(ii) 'John believes that D′'.

Suppose we examine John with the help of a comprehensive list of sentences which are L-true in *S*; among them, for instance, are translations into English of theorems in the system of [P.M.] and of even more complicated mathematical theorems which can be proved in that system and therefore are L-true on the basis of the accepted interpretation. We ask

John, for every sentence or for its negation, whether he believes what it says or not. Since we know him to be truthful, we take his affirmative or negative answer as evidence for his belief or nonbelief. Among the simple L-true sentences, there will certainly be some for which John professes belief. We take as 'D' any one of them, say 'Scott is either human or not human'. Thus the sentence (i) is true. On the other hand, since John is a creature with limited abilities, we shall find some L-true sentences in *S* for which John cannot profess belief. This does not necessarily mean that he commits the error of believing their negations; it may be that he cannot give an answer either way. We take as 'D′' some sentence of this kind; that is to say, 'D′' is L-true but (ii) is false. Thus the two belief-sentences (i) and (ii) have different truth-values; they are neither equivalent nor L-equivalent. Therefore, the definitions of interchangeability and L-interchangeability (11-1a) lead to the following two results:

13-1. The occurrence of 'D' within (i) is not interchangeable with 'D′'.
13-2. The occurrence of 'D' within (i) is not L-interchangeable with 'D′'.

'D' and 'D′' are both L-true; therefore:

13-3. 'D' and 'D′' are equivalent and L-equivalent.

Examining the first belief-sentence (i) with respect to its subsentence 'D', we see from 13-1 and 13-3 that the condition of extensionality (11-2a) is not fulfilled; and we see from 13-2 and 13-3 that the condition of intensionality (11-3a) is not fulfilled either:

13-4. The belief-sentence (i) is *neither extensional nor intensional* with respect to its subsentence 'D'.

Although 'D' and 'D′' have the same intension, namely, the L-true or necessary proposition, and hence the same extension, namely, the truth-value truth, their interchange transforms the first belief-sentence (i) into the second (ii), which does not have the same extension, let alone the same intension, as the first.

The same result as 13-4 holds also if any other sentence is taken instead of 'D', in particular, any factual sentence.

Let us now try to answer the much-discussed question as to how a sentence reporting a belief is to be analyzed and, in particular, whether such a sentence is about a proposition or a sentence or something else. It seems to me that we may say, in a certain sense, that (i) is about the sentence 'D', but also, in a certain other sense, that (i) is about the proposition that D. In interpreting (i) with respect to the sentence 'D', it would, of course, not do to transform it into 'John is disposed to an affirmative response to the sentence 'D' ', because this might be false, although (i) was assumed to be true; it might, for instance, be that John does not understand English but expresses his belief in another language. Therefore, we may try the following more cautious formulation:

(iii) 'John is disposed to an affirmative response to some sentence in some language, which is L-equivalent to 'D' '.

Analogously, in interpreting (i) with respect to the proposition that D, the formulation 'John is disposed to an affirmative response to any sentence expressing the proposition that D' would be wrong because it implies that John understands all languages. Even if the statement is restricted to sentences of the language or languages which John understands, it would still be wrong, because 'D″', for example, or any translation of it, likewise expresses the proposition that D, but John does not give an affirmative response to it. Thus we see that here again we have to use a more cautious formulation similar to (iii):

(iv) 'John is disposed to an affirmative response to some sentence in some language which expresses the proposition that D'.

However, it seems to me that even the formulations (iii) and (iv), which are L-equivalent, should not be regarded as anything more than a first approximation to a correct interpretation of the belief-sentence (i). It is true that each of them follows from (i), at least if we take 'belief' here in the sense of 'expressible belief', leaving aside the problem of belief in a wider sense, interesting though it may be. However, (i) does not follow from either of them. This is easily seen if we replace 'D' by 'D″'. Then (iii) remains true because of 13-3; on the other hand, (i) becomes (ii), which is false. It is clear that we must interpret (i) as saying as much as (iii) but something more; and this additional content seems difficult to formulate. If (i) is correctly interpreted in accord with its customary meaning, then it follows from (i) that there is a sentence to which John would respond affirmatively and which is not only L-equivalent to 'D', as (iii) says, but has a still stronger relation to 'D'—in other words, a sentence which has something more in common with 'D' than the intension. The two sentences must, so to speak, be understood in the same way; they must not only be L-equivalent in the whole but consist of L-equivalent parts, and both must be built up out of these parts in the same way. If this is the case, we shall say that the two sentences have the same intensional structure. This concept will be explicated in the next section and applied in the analysis of belief sentences in § 15.

§ 14. Intensional Structure

If two sentences are built in the same way out of designators (or designator matrices) such that any two corresponding designators are L-equivalent, then we say that the two sentences are *intensionally isomorphic* or that they have the same *intensional structure*. The concept of L-equivalence can also be used in a wider sense for designators in different language systems; and the concept of intensional isomorphism can then be similarly extended.

We shall discuss here what we call the analysis of the intensional structures of designators, especially sentences. This is meant as a semantical analysis, made on the basis of the semantical rules and aimed at showing, say for a given sentence, in which way it is built up out of designators and what are the intensions of these designators. If two sentences are built in the same way out of corresponding designators with the same intensions, then we shall say that they have the same intensional structure. We might perhaps also use for this relation the term 'synonymous', because it is used in a similar sense by other authors (e.g., Langford, Quine, and Lewis), as we shall see in the next section. We shall now try to explicate this concept.

Let us consider, as an example, the expressions '2 + 5' and 'II sum V' in a language *S* containing numerical expressions and arithmetical functors. Let us suppose that we see from the semantical rules of *S* that both '+' and 'sum' are functors for the function Sum and hence are L-equivalent; and, further, that the numerical signs occurring have their ordinary meanings and hence '2' and 'II' are L-equivalent to one another, and likewise '5' and 'V'. Then we shall say that the two expressions are *intensionally isomorphic* or that they have *the same intensional structure*, because they not only are L-equivalent as a whole, both being L-equivalent to '7', but consist of three parts in such a way that corresponding parts are L-equivalent to one another and hence have the same intension. Now it seems advisable to apply the concept of intensional isomorphism in a somewhat wider sense so that it also holds between expressions like '2 + 5' and 'sum(II,V)', because the use in the second expression of a functor preceding the two argument signs instead of one standing between them or of parentheses and a comma may be regarded as an inessential syntactical device. Analogously, if '>' and 'Gr' are L-equivalent, and likewise '3' and 'III', then we regard '5 > 3' as intensionally isomorphic to 'Gr(V,III)'. Here again we regard the two predicators '>' and 'Gr' as corresponding to each other, irrespective of their places in the sentences; further, we correlate the first argument expression of '>' with the first of 'Gr', and the second with the second. Further, '2 + 5 > 3' is isomorphic to 'Gr[sum(II,V),III]', because the corresponding expressions '2 + 5' and 'sum(II,V)' are not only L-equivalent but isomorphic. On the other hand, '7 > 3' and 'Gr[sum(II,V),III]' are not isomorphic; it is true that here again the two predicators '>' and 'Gr' are L-equivalent and that corresponding argument expressions of them are likewise L-equivalent, but the corresponding expressions '7' and 'sum(II,V) are not isomorphic. We require for isomorphism of two expressions that the analysis of both down to the smallest subdesignators lead to analogous results.

We have said earlier (§ 1) that it seems convenient to take as designators in a system *S* at least all those expressions in *S*, but not necessarily only those, for which there are corresponding variables in the metalanguage *M*. For the present purpose, the comparison of intensional structures, it seems advisable to go as far as possible and take as designators all those expressions which serve as sentences, predicators, functors, or individual expressions of any type, irrespective of the question of whether or not *M* contains corresponding variables. Thus, for example, we certainly want to regard as isomorphic '$p \vee q$' and 'Apq', where 'A' is the sign of disjunction (or alternation) as used by the Polish logicians in their parenthesis-free notation, even if *M*, as is usual, does not contain variables of the type of connectives. We shall then regard '$\vee$' and 'A' as L-equivalent connectives because any two full sentences of them with the same argument expressions are L-equivalent.

Frequently, we want to compare the intensional structures of two expressions which belong to different language systems. This is easily possible if the concept of L-equivalence is defined for the expressions of both languages in such a way that the following requirement is fulfilled, in analogy to our earlier conventions: an expression in *S* is L-equivalent to an expression in *S'* if and only if the semantical rules of *S* and *S'* together, without the use of any knowledge about (extra-linguistic) facts, suffice to show that the two expressions have the same extension. Thus, L-equivalence holds, for example, between 'a' in *S* and 'a'' in *S'* if we see from the rules of designation for these two individual constants that both stand for the same individual; likewise between 'P' and 'P'', if we see from the rules alone that these predicators apply to the same individuals; between two functors '+' and 'sum', if we see from the rules alone that they assign to the same arguments the same values—in other words, if their full expressions with L-equivalent argument expressions (e.g., '2 + 5' and 'sum(II,V)') are L-equivalent; for two sentences, if we see from the rules alone that they have the same truth-value (e.g., 'Rom ist gross' in

German, and 'Rome is large' in English). Thus, even if the sentences '$2 + 5 > 3$' and 'Gr [sum(II,V),III]' belong to two different systems, we find that they are intensionally isomorphic by establishing the L-equivalence of corresponding signs.

If variables occur, the analysis becomes somewhat more complicated, but the concept of isomorphism can still be defined. We shall not give here exact definitions but merely indicate, with the help of some simple examples, the method to be applied in the definitions of L-equivalence and isomorphism of matrices. Let 'x' be a variable in S which can occur in a universal quantifier '(x)' and also in an abstraction operator '(λx)', and 'u' be a variable in S' which can occur in a universal quantifier 'Πu' and also in an abstraction operator '$\hat{u}$'. If 'x' and 'u' have the same range of values (or, more exactly, of value intensions, § 10), for example, if both are natural number variables (have natural number concepts as value intensions), we shall say that 'x' and 'u' are L-equivalent, and also that '(x)' and 'Πu' are L-equivalent, and that '(λx)' and '$\hat{u}$' are L-equivalent. If two matrices (sentential or other) of degree n are given, one in S and the other in S', we say that they are L-equivalent with respect to a certain correlation between the variables, if corresponding abstraction expressions are L-equivalent predicators. Thus, for example, '$x > y$' in S and 'Gr(u,v)' in S' are L-equivalent matrices (with respect to the correlation of 'x' with 'u' and 'y' with 'v') because '(λxy) [$x > y$]' and '$\hat{u}\hat{v}$[Gr(u,v)]' are L-equivalent predicators. Intensional isomorphism of (sentential or other) matrices can then be defined in analogy to that of closed designators, so that it holds if the two matrices are built up in the same way out of corresponding expressions which are either L-equivalent designators or L-equivalent matrices. Thus, for example, the matrices '$x + 5 > y$' and 'Gr[sum(u,V),v]' are not only L-equivalent but also intensionally isomorphic; and so are the (L-false) sentences '$(x)(y)$[$x + 5 > y$]' and '$\Pi u \Pi v$ [Gr[sum(u,V),v]]'.

These considerations suggest the following definition, which is recursive with respect to the construction of compound designator matrices out of simpler ones. It is formulated in general terms with respect to designator matrices; these include closed designators and variables as special cases. The definition presupposes an extended use of the term 'L-equivalent' with respect to variables, matrices, and operators, which has been indicated in the previous examples but not formally defined. The present definition makes no claim to exactness; an exact definition would have to refer to one or two semantical systems whose rules are stated completely.

14-1. *Definition of intensional isomorphism*

a. Let two designator matrices be given, either in the same or in two different semantical systems, such that neither of them contains another designator matrix as proper part. They are intensionally isomorphic $=_{Df}$ they are L-equivalent.

b. Let two compound designator matrices be given, each of them consisting of one main submatrix (of the type of a predicator, functor, or connective) and *n* argument expressions (and possibly auxiliary signs like parentheses, commas, etc.). The two matrices are intensionally isomorphic $=_{Df}$ (1) the two main submatrices are intensionally isomorphic, and (2) for any *m* from 1 to *n*, the *m*th argument expression within the first matrix is intensionally isomorphic to the *m*th in the second matrix ('the *m*th' refers to the order in which the argument expressions occur in the matrix).

c. Let two compound designator matrices be given, each of them consisting of an operator (universal or existential quantifier, abstraction operator, or description operator) and its scope, which is a designator matrix. The two matrices are intensionally isomorphic $=_{Df}$ (1) the two scopes are intensionally isomorphic with respect to a certain correlation of the variables occurring in them, (2) the two operators are L-equivalent and contain correlated variables.

In accord with our previous discussion of the explicandum, rule b in this definition takes into consideration the order in which argument expressions occur but disregards the place of the main subdesignator. For the intensional structure, in contrast to the merely syntactical structure, only the order of application is essential, not the order and manner of spelling.

§ 15. Applications of the Concept of Intensional Structure

The concept of intensional structure is compared with the concepts of synonymity discussed by Quine and Lewis. The concept is then used for giving an interpretation of belief sentences that seems more adequate than the interpretations discussed earlier (§ 13). Further, the same concept helps in solving the so-called paradox of analysis.

It has often been noticed by logicians that for the explication of certain customary concepts a stronger meaning relation than identity of intension seems to be required. But usually this stronger relation is not defined. It seems that in many of these cases the relation of intensional isomorphism could be used. For example, if we ask for an exact translation of a given statement, say the exact translation of a scientific hypothesis or of the testimony of a witness in court from French into English, we should usually require much more than agreement in the intensions of the sentences, that is, L-equivalence of the sentences. Even if we restrict our attention to designative (cognitive) meaning—leaving aside other meaning components like the emotive and the motivative, although they are often very important even for the translation of theoretical texts—L-equivalence of sentences is not sufficient; it will be required that at least some of the component designators be L-equivalent, in other words, that the intensional structures be alike or at least similar.

Quine explains, without giving a definition, a concept of synonymity which is different from and presumably stronger than L-equivalence. He says: "The notion of synonymity figures implicitly also whenever we use the method of indirect quotations. In indirect quotation we do not insist on a literal repetition of the words of the person quoted, but we insist on a synonymous sentence; we require reproduction of the *meaning*. Such synonymity differs even from logical equivalence; and exactly what it is remains unspecified."[38] We might perhaps think of an explicatum of this concept of synonymity similar to our concept of intensional isomorphism. Quine himself seems to expect that the explication will be found not in semantics but in what we would call pragmatics, because he says that the concept of synonymity "calls for a definition or a criterion in psychological and linguistic terms."

C. I. Lewis[39] gives a definition for the concept of synonymity which shows a striking similarity to our concept of intensional isomorphism, although the two concepts have been developed independently. Since it is interesting to see the points of agreement and of difference, I will quote his explanations at length. "Not every pair of expressions having the same intension would be called synonymous; and there is good reason for this fact. Two expressions are commonly said to be synonymous (or in the case of propositions, equipollent) if they have the same intension, and *that intension is neither zero nor universal*. But to say that two expressions with the same intension have the same meaning, without qualification, would have the anomalous consequence that any two analytic propositions would then be equipollent, and any two self-contradictory propositions would be equipollent." In order to overcome this difficulty, Lewis introduces a new concept: "Two expressions are *equivalent in analytic meaning*, (1) if at least one is elementary [i.e., not complex] and they have the same intension, or (2) if, both being complex, they can be so analyzed into constituents that

[38] [Notes], p. 120.

[39] [Meaning], pp. 245 f. Other concepts used by Lewis will be discussed in the next section.

(*a*) for every constituent distinguished in either, there is a corresponding constituent in the other which has the same intension, (*b*) no constituent distinguished in either has zero intension or universal intension, and (*c*) the order of corresponding constituents is the same in both, or can be made the same without alteration of the intension of either whole expression." As examples, Lewis states that "round excision" and "circular hole" are equivalent in analytic meaning, while "equilateral triangle" and "equiangular triangle" are not, although they have the same intension. He continues: "We shall be in conformity with good usage if we say that two expressions are synonymous or equipollent, (1) if they have the same intension and that intension is neither zero nor universal, or (2) if, their intension being either zero or universal, they are equivalent in analytic meaning."

Thus Lewis' concept of synonymity is very similar to our concept of intensional isomorphism except for one point: He applies this stronger relation only to the two extreme cases of intension, for example, in the field of sentences, only to L-determinate and not to factual sentences. This discrimination seems to me somewhat arbitrary and inadvisable. Let us consider the following examples (in a language which, in distinction to S_1, also contains expressions for finite cardinal numbers and for relations and properties of them):

(i) 'two is an even prime number';
(ii) 'two is between one and three';
(iii) 'the number of books on this table is an even prime number';
(iv) 'the number of books on this table is between one and three'.

The sentences (i) and (ii) have the same intension but are not equivalent in analytic meaning (intensionally isomorphic). The same holds for (iii) and (iv). Now, according to Lewis' definition, (i) and (ii) are not synonymous because they are L-true, analytic; while (iii) and (iv) are synonymous because they are factual, synthetic. It seems to me that it would be more natural to regard (iii) and (iv) also as nonsynonymous, since the difference between them is essentially the same as that between (i) and (ii). The logical operation which leads from (i) to (ii) is the same as that which leads from (iii) to (iv); it is the transformation of '*n* is an even prime number' into '*n* is (a cardinal number) between one and three'.

Now let us go back to the problem of the analysis of belief-sentences, and let us see how the concept of intensional structure can be utilized there. It seems that the sentence 'John believes that D' in *S* can be interpreted by the following semantical sentence:

15-1. 'There is a sentence $\mathfrak{S}_i$ in a semantical system S' such that (*a*) $\mathfrak{S}_i$ in S' is intensionally isomorphic to 'D' in S and (*b*) John is disposed to an affirmative response to $\mathfrak{S}_i$ as a sentence of S'.'

This interpretation may not yet be final, but it represents a better approximation than the interpretations discussed earlier (in § 13). As an example, suppose that John understands only German and that he responds affirmatively to the German sentence 'Die Anzahl der Einwohner von Chicago ist grösser als 3,000,000' but neither to the sentence 'Die Anzahl der Einwohner von Chicago ist grösser als $2^6 \times 3 \times 5^6$' nor to any intensionally isomorphic sentence, because he is not quick enough to realize that the second sentence is L-equivalent to the first. Then our interpretation of belief-sentences, as formulated in 15-1, allows us to assert the sentence 'John believes that the number of inhabitants of Chicago is greater than three million' and to deny the sentence 'John believes that the number of inhabitants of Chicago is greater than $2^6 \times 3 \times 5^6$'. We can do so without contradiction because the two German sentences, and likewise their English translations just used, have different intensional structures. [By the way, this example shows another disadvantage of Lewis' definition of equivalence in analytic meaning. According to part (1) of his definition, the two German sentences are equivalent in analytic meaning if we take '3,000,000' as one sign.] On the other hand, the interpretation of belief-sentences in terms of propositions as objects of beliefs (like (iv) in § 13) would not be adequate in this case, since the two German sentences and the two English sentences all express the same proposition.

An analogous interpretation holds for other sentences containing psychological terms about knowledge, doubt, hope, fear, astonishment, etc., with 'that'-clauses, hence generally about what Russell calls propositional attitudes and Ducasse epistemic attitudes. The problem of the logical analysis of sentences of this kind has been much discussed,[40] but a satisfactory solution has not been found so far. The analysis here proposed is not yet a complete solution, but it may perhaps be regarded as a first step. What remains to be done is, first, a refinement of the analysis in terms of linguistic reactions here given and, further, an analysis in terms of dispositions to nonlinguistic behavior.

[40] Russell, [Inquiry], gives a detailed discussion of the problem in a wider sense, including beliefs not expressed in language; he investigates the problem under both an epistemological and a logical aspect (in our terminology, both a pragmatical and a semantical aspect), not always distinguishing the two clearly. For C. J. Ducasse's conception see his paper "Propositions, Opinions, Sentences, and Facts," *Journal of Philosophy*, XXXVII (1940), 701–11.

The concept of intensional structure may also help in clarifying a puzzling situation that has been called "the paradox of analysis". It was recently stated by G. E. Moore,[41] and then discussed by C. H. Langford,[42] Max Black,[43] and Morton White.[44] Langford[45] states the paradox as follows: "If the verbal expression representing the analysandum has the same meaning as the verbal expression representing the analysans, the analysis states a bare identity and is trivial; but if the two verbal expressions do not have the same meaning, the analysis is incorrect." Consider the following two sentences:

'The concept Brother is identical with the concept Male Sibling.'
'The concept Brother is identical with the concept Brother.'

The first is a sentence conveying fruitful information, although of a logical, not a factual, nature; it states the result of an analysis of the analysandum, the concept Brother. The second sentence, on the other hand, is quite trivial. Now Moore had been puzzled by the following fact: If the first sentence is true, then the second seems to make the same statement as the first (presumably because, if two concepts are identical, then a reference to the one means the same as a reference to the other, and hence the one expression can be replaced by the other); "but it is obvious that these two statements are not the same", he says. Black tries to show that the two sentences do not express the same proposition; he supports this assertion by pointing to the fact that the first sentence, or rather a paraphrasing he gives for it ('the concept Brother is the conjunct of the concept Male and the concept Sibling') refers to a certain non-identical relation (the triadic relation Conjunct), while the second is a mere identity. White replies that this is not a sufficient reason for the assertion. None of the four authors states his criterion for the identity of "meaning", "statement", or "proposition"; this seems the chief cause for the inconclusiveness of the whole discussion. If we take, as in the terminology used in this book, L-equivalence as the condition for the identity of propositions, then White is certainly right; since the two sentences are L-true and hence L-equivalent to each other, they express the same proposition in our sense. On the other hand, Black feels correctly, like Moore and Langford, that there is an important difference in meaning between the two sentences, because of a difference in meaning between

[41] *The Philosophy of G. E. Moore*, ed. P. Schilpp (1942), pp. 660–67.

[42] "The Notion of Analysis in Moore's Philosophy", *ibid.*, pp. 321–42.

[43] *Mind*, LIII (1944), 263–67 and LIV (1945), 272 f.

[44] *Mind*, LIV (1945), 71 f. and 357–61.

[45] *Op. cit.*, p. 323.

the two expressions for the analysandum ('the concept Brother') and the analysans ('the concept Male Sibling'). The paradox can be solved if we can state exactly what this difference in meaning is and how it is compatible with the identity of meaning in another sense. The solution is quite simple in terms of our concepts: The difference between the two expressions, and, consequently, between the two sentences is a difference in intensional structure, which exists in spite of the identity of intension. Langford saw the point at which the difference lies; he says[46] that the analysans is more articulate than the analysandum, it is a grammatical function of more than one idea; the two expressions are not synonymous but "cognitively equivalent in some appropriate sense". It seems to me that this cognitive equivalence is explicated by our concept of L-equivalence and that the synonymity, which does not hold for these expressions, is explicated by intensional isomorphism.

§ 16. Lewis' Method of Meaning Analysis

Lewis uses, in addition to the concepts of extension and intension which are similar to ours, the concept of comprehension which presupposes the admission of nonactual, possible things. It seems inadvisable to use this conception because it requires a new, more complicated language form. The distinction which Lewis wants to make can better be made with respect to intensions than with respect to things.

I wish to discuss briefly some concepts which have recently been proposed by C. I. Lewis[47] as tools for a semantical meaning analysis. There is a striking similarity between these concepts and our concepts of extension and intension. This similarity is due to the common aim to make some traditional concepts, especially extension and intension, denotation and connotation, more general in their application and, at the same time, more clear and precise.

Lewis explains his chief semantical concepts in the following way: "All terms have meaning in the sense or mode of denotation or extension; and all have meaning in the mode of connotation or intension. The *denotation* of a term is the class of all actual or existent things to which that term correctly applies. . . . The *comprehension* of a term is the classifica-

[46] *Op. cit.*, p. 326.

[47] In [Meaning]. This paper is part of a "Symposium on Meaning and Truth", published in four parts in *Philosophy and Phenomenological Research*, Vols. IV (1943–44) and V (1944–45). This symposium also contains a number of other interesting contributions to the development and clarification of semantical concepts. I have elsewhere referred to Tarski's paper [Truth]; I am in close agreement with his conception of the nature of semantics, but he does not discuss the central problems of this book. Concerning these problems, I wish especially to call attention to the papers by C. J. Ducasse (IV, 317–40; V, 320–32) and Charles A. Baylis (V, 80–93).

tion of all consistently thinkable things to which the term would correctly apply. . . . For example, the comprehension of "square" includes all imaginable as well as all actual squares, but does not include round squares. . . . The *connotation* or *intension* of a term is delimited by any correct definition of it."

It seems that Lewis' concepts of extension and intension correspond closely to our concepts. This is clearly the case for predicators, but perhaps also for sentences and individual expressions. There remains the problem of the necessity and usefulness of Lewis' third concept, that of comprehension. It seems that Lewis follows Meinong[48] in dividing (1) all things (in the widest sense) into impossible or inconceivable things (e.g., round squares) and possible things; and (2) the possible things into actual things (e.g., Plato) and nonactual possible things (e.g., Apollo, unicorns). [Lewis clearly makes the second division. Whether he also makes the first and hence countenances, like Meinong, impossible things is not quite so clear but seems indicated by the formulation that the comprehension "does not include round squares". According to the ordinary conception, in distinction to Meinong's, there are no round squares at all, not even in some particular kind of objects; hence it would be redundant to say that the comprehension "does not include round squares".] Meinong's conception has been critically discussed by Russell[49] and then rejected. Russell's chief reason for the rejection is that the impossible objects violate the principle of contradiction; for example, a round square is both round and nonround, because square. Russell is certainly right in the following respect: Within the logical framework of our ordinary language, we cannot consistently apply the conception of impossible things or even that of possible nonactual things. And, as far as I am aware, neither Meinong nor Lewis nor any other philosopher has constructed or even outlined a language of a new structure which would accommodate those entities. That such a language must be different from the ordinary one is shown by the following example: In the ordinary language we say: 'There are no white ravens and no round squares'. In the new language we would have to say, instead: 'There are white ravens; however, they are not actual, but only possible. And there are round squares; however, they are neither actual nor possible, but impossible.' I have no doubt that a resourceful logician could easily construct a consistent language system of this kind, if we wanted it; he would have to lay down rules for the quantifiers deviating from the ordinary rules in a way suggested by the examples. The

[48] A. von Meinong, *Untersuchungen zur Gegenstandstheorie und Psychologie* (1904).

[49] [Denoting], pp. 482 f.

decisive question is not that of the technical possibility of such a language but rather that of its usefulness. Only if it can be shown to have great advantages in comparison to the ordinary language structure would it be worth considering in spite of its fundamental deviation and increased complexity.

I do not see sufficient reasons for this change. The distinctions which Meinong and Lewis have in mind are important, but they can be taken care of in a different way. Instead of dividing objects into (i) actual, (ii) nonactual but possible, and (iii) impossible, we make analogous distinctions, first, between three corresponding kinds of expressions and then between three corresponding kinds of intensions. Let us show this, first, for predicators. Instead of speaking about three kinds of objects like this:

(i) '(some) horses are actual objects',
(ii) 'unicorns are nonactual but possible objects',
(iii) 'round squares are impossible objects',

we speak, rather, about three kinds of predicators:

(i) 'the predicator 'horse' is not empty',
(ii) 'the predicator 'unicorn' is F-empty, i.e., empty but not L-empty',
(iii) 'the predicator 'round square' is L-empty'.

Then we apply the same terms to the corresponding intensions (this is a transference of terms from a semantical to a nonsemantical use, analogous to the transference of the terms 'equivalent' and 'L-equivalent', § 5):

(i) 'the property Horse is not empty',
(ii) 'the property Unicorn is F-empty, i.e., empty but not L-empty',
(iii) 'the property Round Square is L-empty'.

An analogous distinction can be made for individual expressions, for instance, descriptions. (We apply here, not the special interpretation of descriptions which we adopted in § 8 because of its technical advantages, but the customary interpretation, according to which a description has a descriptum only if the uniqueness condition is fulfilled.) Then, instead of using the following formulations referring to objects:

(i) 'Alexander's horse (i.e., the one horse which Alexander had at such and such a time) is an actual object',
(ii) 'Alexander's unicorn is a nonactual but possible object',
(iii) 'Alexander's round square is an impossible object',

we use, rather, the following ones concerning individual expressions (Lewis' singular terms):

(i) 'the description 'Alexander's horse' is not empty',
(ii) 'the description 'Alexander's unicorn' is F-empty' (in Lewis' terminology, it has zero denotation, but not zero comprehension);
(iii) 'the description 'Alexander's round square' is L-empty' (it has zero comprehension).

And then we make analogous statements concerning the corresponding individual concepts (in Lewis' terminology, connotations of singular terms):

(i) 'the individual concept Alexander's Horse is not empty',
(ii) 'the individual concept Alexander's Unicorn is F-empty',
(iii) 'the individual concept Alexander's Round Square is L-empty'.

Thus our method does not neglect the distinctions pointed out by Meinong and Lewis. However, it applies the distinction to intensions, while these philosophers apply it to objects and thereby violate the rule of ordinary language which takes the addition of 'actual' to a general noun as redundant. For example, the ordinary language takes phrases like 'actual horses', 'real horses', 'existing horses', etc. (where 'actual', etc., does not mean 'occurring at the present time' but 'occurring at some time, past, present, or future'), as meaning the same as 'horses', differing from this only in emphasis; and, likewise, 'actual unicorns' is taken as meaning the same as 'unicorns', and hence it is said: 'there are no unicorns (at any space-time point)'.

If we thus reject such distinctions between kinds of objects, then Lewis' concept of comprehension can no longer be defined. Do we hereby sacrifice a useful tool of semantical meaning analysis? I do not think so. Lewis emphasizes rightly the difference between comprehension and extension. But there seems not to be much difference between the purposes of the concepts of comprehension and *intension*. If we accept Lewis' language form, then these concepts are both legitimate and, of course, not identical. But whatever is said in terms of comprehension can immediately be translated into terms of intension, because comprehension and intension determine each other logically. If you tell me the comprehension of a Chinese word, then I know immediately what is its intension, and vice versa; therefore there is no advantage in having both concepts. On the other hand, if you tell me the intension of a Chinese word, I do not know its extension (unless it is L-determinate); and if you tell me only its extension, I cannot infer from this its intension. Therefore, it is useful to have both concepts, that of intension and that of extension.

We also arrive at the same result, the rejection of nonactual, possible objects and of comprehension by an approach from another angle, that of modal logic. We shall find later (§§ 42 f.) that the logical modalities must be applied to intensions, not to extensions. Thus we may speak of an impossible (or L-false) proposition but not of an impossible truth-value; of an impossible (or L-empty) property but not of an impossible (or L-empty) class. Analogously, we may speak of an impossible (or L-empty) individual concept but not of an impossible individual (object, thing), because individuals (objects, things) are extensions, not intensions; in other words, individuals are involved in questions of application, not in questions of meaning in the strict sense. (We take here, of course, the ordinary conception of extensions, not that to be discussed in § 23, according to which extensions are construed as a special kind of intension.)

To sum up, I do not think that the concepts of possible and impossible objects and of comprehension can be accused of violating logic or of leading necessarily to contradictions. However, it seems doubtful whether these concepts are sufficiently useful to compensate for their disadvantage —the necessity of using an uncustomary and more complex language structure.

CHAPTER II

L-DETERMINACY

We have seen (§ 2) that a sentence is L-determinate if its truth-value, which is its extension, is determined by the semantical rules. In this chapter we apply the concept of L-determinacy also to other designators. The definitions are constructed so that an analogous result holds: A designator is L-determinate if the semantical rules, independently of facts, suffice for determining its extension (§ 17). For the application of this concept we presuppose that the individuals are positions in an ordered domain. An individual expression is L-determinate if the semantical rules suffice for determining the location of the position to which it refers (§§ 18, 19). A predicator is L-determinate if the semantical rules suffice for determining for every position whether the predicator applies to it or not (§ 20). The distinction between logical and descriptive (nonlogical) signs is discussed, and its connection with the distinction between L-determinate and L-indeterminate designators is examined (§ 21). The intension of an L-determinate designator is also called L-determinate (§ 22). There is a one-one correlation between extensions and L-determinate intensions; therefore, it would be possible, though not customary, to define extensions as L-determinate intensions (§ 23).

§ 17. L-Determinate Designators

In general, factual knowledge is needed for establishing the truth-value of a given sentence. However, if the sentence is L-determinate (§ 2), the semantical rules suffice for establishing its truth-value or, in other words, its extension. The concept of L-determinacy will now be extended to designators of other kinds. We stipulate that the definitions of this concept for the other kinds be such that a designator is L-determinate if and only if the semantical rules suffice for determining its extension. Definitions fulfilling this requirement will be constructed in later sections of this chapter.

We found earlier that the intension of the sentence 'Hs' in the system S_1 is the proposition that Scott is human and that its extension is the truth-value truth. Now let us consider the question of what knowledge we need in this and other cases in order to determine the intension and the extension of a given sentence. It is clear that, for the determination of the intension, only the semantical rules of the system S_1 are required. For every sentence in S_1 these rules give an interpretation and thereby tell us what proposition is the intension of the sentence. Thus the result mentioned concerning the intension of 'Hs' is established on the basis of those rules which give an interpretation for 'Hs'; these are the rules of designation for 'H' and for 's' (see 1-1 and 1-2) and the rule of truth for atomic sentences (1-3). On the other hand, for the determination of the extension, the truth-value, of 'Hs', knowledge of the semantical rules alone is obviously not sufficient. We need, in addition, factual knowledge. This factual knowledge is based on observations of the thing Walter Scott; these observations lead to the result that this thing has the properties characteristic of human beings and, hence, that the sentence 'Hs' is true.

However, we have seen that there is a particular kind of sentence for the determination of whose truth-values the semantical rules without any factual knowledge provide a sufficient basis. These are the L-determinate sentences, that is, the L-true and the L-false sentences (see the explanation preceding 2-7). Thus, for these sentences the semantical rules suffice to determine not only their intensions but also their extensions. Now we shall extend the meaning of the term 'L-determinate' so as to make it applicable to designators in general, in analogy to its application to sentences. For this purpose it seems natural to lay down the following convention for any semantical system *S:*

17-1. A designator is *L-determinate* in *S* if and only if its extension can be determined on the basis of the semantical rules of *S* alone, without any reference to facts.

This convention is not itself a definition of 'L-determinate'. It is meant merely as an informal characterization of the explicandum; in other words, a requirement which the definition should fulfil. A definition of L-determinacy for sentences has already been given (2-3d). The problems of constructing definitions of L-determinacy for other kinds of designators will be discussed in the subsequent sections. But even if it is regarded as merely a requirement, the present formulation in 17-1 is found upon examination to be insufficient. The phrase "the extension is determined by certain rules" can be understood in two quite different senses. We have to find out which sense is appropriate here.

The difficulty here involved can perhaps best be made clear in the case of a predicator. The intension of the predicator 'H' can obviously be determined with the help of the semantical rules alone; we see from the rule of designation for 'H' (1-2) that its intension is the property Human. But does the same not hold for the extension, too? Do we not also see from the same rule that the extension of 'H' is the class Human? Should we then say, according to our convention, that 'H', and likewise every other predicator, is L-determinate? This would obviously not be in accordance with the intended meaning of this term.

In order to overcome this difficulty, we have to make a certain distinction which can easily be explained for sentences and then transferred to designators of other kinds. Suppose we ask the question: "What is the extension, that is, the truth-value, of the sentence 'Hs'?" Consider the following sentences under 17-2 and 17-3, which belong to the metalanguage M. Let us examine whether they may be regarded as satisfactory answers to our question.

17-2. **a.** 'The extension of 'Hs' is the truth-value truth.'
b. ''Hs' is true.'
c. 'Scott is human.'
d. 'The extension of 'Hs' is the same as that of 'H ≡ H'.'
e. ''Hs' is equivalent to 'H ≡ H'.'

17-3. **a.** 'The extension of 'Hs' is the truth-value that Scott is human.'
b. ''Hs' is true if and only if Scott is human.'

Each of these seven sentences is true (see 6-3). And in some sense each of them may be said to give an answer to our question. However, there is an important difference between the sentences under 17-2 and those under 17-3. Suppose we understand the sentences of the system S_1 but have no factual knowledge concerning the things referred to in these sentences; then we do not know whether 'Hs' is true or not, in other words, whether Scott is human or not. Suppose, further, that the purpose of our question was to find this lacking knowledge. Then 17-2a is a completely satisfactory answer because it supplies the information we want; and so is 2b, which is merely a simpler formulation for 2a; and likewise 2c, which gives the same information without the use of semantical terms. (For the result that 2b and 2c mean the same, see the explanation preceding 1-7.) On the other hand, the answer 3a, although correct, does not satisfy our purpose; we shall reply with a modified formulation of our first question: "Yes; but what *is* the truth-value that Scott is human?" Similarly, we shall reply to 3b: "Yes; but *is* Scott human or not?" We may formulate this difference by saying that 2a, 2b, and 2c actually *give* the truth-value of 'Hs', while 3a and 3b do not give it but merely *describe* it, in the sense of supplying a description for it (in Russell's sense of 'description'). We can do this by introducing the phrase 'gives the truth-value' in the following way, which is not meant as an exact definition. Let $\mathfrak{S}_j$ be a true sentence in M (it may also be a definition or a rule or a set of true sentences, definitions, or rules). We shall say that the truth-value of a sentence $\mathfrak{S}_i$ in a system *S* is *given* by $\mathfrak{S}_j$ if either the sentence '$\mathfrak{S}_i$ is true (in *S*)' or its negation follows from $\mathfrak{S}_j$ (in M) without the use of any factual knowledge not supplied by $\mathfrak{S}_j$. [The phrase '. . . follows from - - - (in M)' may be understood as meaning the same as '. . . is L-implied by - - - (in M)', if we assume that L-terms with respect to M have been defined in the metametalanguage

MM. For the sake of simplicity, we use the German letters with subscripts not only in M for expressions in *S* but also in MM for expressions in M and expressions in *S*.]

Let us now apply this criterion to the sentences under 17-2 and 17-3. First, 2b fulfils the criterion in a trivial way; hence it gives the extension of 'Hs'. Furthermore, each of the sentences 2a and 2c, and even 2d and 2e, gives, together with the semantical rules of S_1, the truth-value of 'Hs', because 2b follows from each of these sentences, together with the rules. That 2b follows from 2a is obvious. Further, 2c follows from 2b, together with the result 3b, which is based on the semantical rules for 'H', 's', and atomic sentences (1-2, 1-1, 1-3). Sentence 2d is derived from 2a and thereby from 2b, together with the result that 'H ≡ H' is L-true, which, in turn, is based on the semantical rules. The same holds for 2e, which is merely another formulation for 2d, according to the definition 5-1. On the other hand, either 3a or 3b, together with the semantical rules, does not give the extension of 'Hs' but merely describes it, because for the derivation of 2b we need here the factual knowledge that Scott is human.

Consider, now, in contrast to 'Hs', an L-determinate sentence, for example, the L-true sentence 's ≡ s' or the L-false sentence '∼(s ≡ s)'. Here no factual sentence $\mathfrak{S}_j$ is required in addition to the semantical rules to give the truth-values of these two sentences. The following two sentences in M follow from the semantical rules of S_1 alone: ''s ≡ s' is true (in S_1)', ''∼(s ≡ s)' is not true (in S_1)'.

In analogy to these results for sentences, we now replace the earlier convention 17-1 by the following:

17-4. A designator is ***L-determinate*** in *S* if and only if the semantical rules of *S* alone, without addition of factual knowledge, *give* its extension.

This again does not yet constitute a definition of 'L-determinate' but only a requirement which the definition should fulfil. For sentences, the previous definition of L-determinacy (2-3d) is in accord with this convention on the basis of our explanation of ". . . gives the extension, i.e., the truth-value, of the sentence - - -". Our task will now be to find adequate definitions of L-determinacy for the other kinds of designators. For each of these kinds we shall have to consider the conditions under which their extensions are actually given, not merely described; as in the case of sentences, "to give the extension" will only be informally explained, not exactly defined. And then the definition of L-determinacy will be constructed in such a way that the requirement 17-4 is fulfilled. If a designator is not L-determinate, we call it ***L-indeterminate.*** This term has been defined for sentences (2-7); however, in the case of sentences, we usually use the synonymous term 'factual'. According to the convention 17-4, a designator is L-indeterminate if its extension can be given only by a factual statement (in M).

§ 18. The Problem of L-Determinacy of Individual Expressions

The conditions under which an individual expression may be regarded as L-determinate are examined. An attempt to base the definition of L-determinacy on a distinction between (genuine) proper names and descriptions is abandoned as inadequate. The analysis is then applied to a coordinate language S_3. Its individuals are positions in a discrete, linear order. 'o', 'o′', 'o″', etc., are the so-called standard individual expressions for these positions in their basic order. Every one of these expressions indicates by its form the location within the basic order to which it refers; hence it exhibits its own extension and may be regarded as L-determinate. The same does not, in general, hold for a description (e.g., 'the one position which is blue and cold'), except when the description is L-equivalent to a standard expression (e.g., 'the one position which is between o′ and o‴').

We begin with individual expressions because, as we shall see later, the solution of the problem of L-determinacy for predicators presupposes the solution for individual expressions.

In analogy to the earlier question, "What is the truth-value of 'Hs'?" we now consider the question, "Which individual is the extension of '$(\imath x)(\mathrm{A}x\mathrm{w})$'?" and possible answers to it. In analogy to the earlier case, let us imagine that we do not know whether there is exactly one author of Waverley and, if so, who he is; and that the purpose of our question is to find out from somebody who does know. Obviously, the answer 'the extension of the description mentioned is the author of Waverley' would not satisfy us even though it is true; it is entirely trivial. [Note that, according to an earlier convention, the phrase 'the author of Waverley' is to be understood as short for 'the one individual who is author of Waverley, or a* if there are no or several such individuals'.] The answer 'the extension sought is the author of Ivanhoe' is true and not trivial; but, nevertheless, it would not satisfy us because it does not supply the specific information we are looking for; we might say here again that this answer merely describes the extension but does not give it. The extension is actually and directly given by the answer 'the extension is Walter Scott'. It is indirectly given by answers like these: 'the extension of '$(\imath x)(\mathrm{A}x\mathrm{w})$' is the same as the extension of 's' ' or ' '$(\imath x)(\mathrm{A}x\mathrm{w}) \equiv \mathrm{s}$' is true'; from these we obtain the direct answer with the help of the semantical rule (1-1), which tells us that the extension of 's' is Walter Scott.

On the basis of these considerations we might perhaps be inclined to propose the following solution: Let us say that the extension of an individual expression is given, and not merely described, by $\mathfrak{S}_j$ if $\mathfrak{S}_j$ uses a proper name in M (e.g., 'Walter Scott') or refers to a proper name in *S* (e.g., 's') in distinction to a description. However, it is easily seen that this does not yet constitute a satisfactory solution. Suppose that '*x* is a dagger and Brutus used *x* for killing Caesar' can be translated into symbols of S_1; this symbolic expression may be indicated by '. . *x* . .'. Suppose somebody gives to our question, 'What is the extension of the description '$(\iota x)(\ldots x \ldots)$'?' the answer 'The extension is the same as the extension of 'b' ', where 'b' is an individual constant in S_1 such that '$(\iota x)(\ldots x \ldots) \equiv$ b' is true. Then the answer is true. According to our tentative solution, we should say that this answer gives the extension of '$(\iota x)(\ldots x \ldots)$', irrespective of the way in which the semantical rule for 'b' is formulated. But suppose, now, that this rule says that 'b' is the symbolic translation of 'the dagger with which Brutus killed Caesar'. Then the above answer says, in other words, that the extension of '$(\iota x)(\ldots x \ldots)$' is the dagger with which Brutus killed Caesar; thus the answer merely describes the extension. The reason for this lies in the fact that the interpretation of the constant 'b' is given in M with the help of a description. We might perhaps say that 'b' is therefore only an apparent proper name, not a genuine one. And we might try to correct the solution proposed by requiring that genuine proper names be used, not those which are defined or interpreted by descriptions. This attempt, however, would lead us into serious difficulties. A moment's reflection shows that most things have no proper names. Some logicians—for example, Russell[1] and Quine[2]—do not accept individual constants as primitive signs but only as abbreviations of compound expressions. Thus the distinction between genuine and apparent proper names of individuals is rather problematic. Even if there are genuine proper names for some individuals, how should the extension of a description be given whose descriptum has no proper name? It is clear that the attempted solution is inadequate in its present form.

However, I believe that another distinction will serve the purpose for which the distinction between proper names and descriptions was intended. To simplify the analysis let us take not a system like S_1, whose individual constants are names of things, but language systems of the following kind. The individuals are positions in an ordered domain. Among the *individual expressions* there are some of a special kind, called

[1] Russell's language contains names for qualities but not for particulars, i.e., individuals in our sense (see [Inquiry], p. 117).

[2] Quine regards all individual constants as abbreviations for descriptions (see [M.L.], pp. 149 ff.).

expressions of ***standard form,*** which fulfil the following condition: (1) if two expressions of standard form are given, then we can see from their forms the positional relation between the two positions. For systems of a simple structure (for example, the system S_3 discussed in this section, in contrast to the language of physics discussed in the next section) the following additional condition is fulfilled: (2) for every position, there is exactly one expression of standard form. Languages of this kind may be called ***coordinate languages,*** in distinction to name languages like S_1.[3] Let us take as an example a language system S_3 in which the basic order of the positions has the simple structure of a progression, a discrete linear order with an initial position but no end. Let 'o' be taken as individual constant for the initial position; if an individual expression of any form, standard or not, is given as an expression for some position, an expression for the next following position is formed from it by the adjunction of a prime '′'. As individual expressions of standard form we take 'o', together with those expressions consisting of 'o' followed by one or several primes. Thus 'o', 'o′', 'o″', 'o‴', are the standard expressions for the first four positions.

Let S_3 contain predicator signs for qualitative properties to be attributed to the positions, say 'B' for the property Blue, 'C' for Cold, 'S' for Soft. Furthermore, S_3 contains, like S_1, the customary connectives, individual variables with quantifiers, and individual descriptions. As common descriptum for all descriptions which do not satisfy the uniqueness condition, we take, of course, the initial position; hence 'o' takes the place of 'a*' (see § 8). Thus, for example, the description '$(\imath x)(Bx \bullet Cx)$' means the same as 'the one position which is both blue and cold (or the position o if no or several positions are both blue and cold)'. [As previously, we shall usually omit the phrase here included in parentheses.]

For the purpose of the subsequent examples, we presuppose this factual assumption:

18-1. The second position (o′) is the only one which is both blue and cold and also the only one which is both blue and soft.

According to this assumption, the following holds:

18-2. '$(\imath x)(Bx \bullet Cx) \equiv o'$' is true (and, moreover, F-true).

18-3. '$(\imath x)(Bx \bullet Sx) \equiv o'$' is true (and, moreover, F-true).

Suppose we ask the question, 'What is the extension of the description '$(\imath x)(Bx \bullet Cx)$'?' because we do not know the facts (18-1) and wish to find

[3] [Syntax], § 3. The system S_3 described in the text is similar to Language I dealt with in [Syntax], Part I.

out which position is the descriptum. Let us consider the following answers:

18-4. a. 'The extension of the description mentioned is the one position which is both blue and cold.'

b. 'The extension of the description mentioned is the same as that of '$(\imath x)(Bx \bullet Sx)$'.'

The answer 18-4a, although true, would certainly appear as unsatisfactory; we should protest: 'Yes, but which position *is* this?'. Sentence 4b is likewise a true answer to our question, by virtue of the fact 18-1. It is not so trivial an answer as 4a, but it still does not supply the information we want. It does not tell us directly which position is the descriptum but merely refers to this position by a qualitative characterization. After receiving the answer 4b, just as in the case of 4a, we still need factual observations concerning the qualities of the positions in order to discover which position is the descriptum of the original description.

In contradistinction to those answers, each of the following two formulations tells us actually what we want to know:

18-5. a. 'The extension of the description is the second position'.

b. 'The extension of the description is the same as that of '$0'$'.'

The same holds for 18-2. Each of these three answers supplies the information directly. But there are other formulations which give the same information in an indirect manner. In order to construct an example, let '$..x..$' indicate a not too simple matrix in S_3 without nonlogical constants, which is fulfilled only by the position $0'$. [We may regard the individual expressions in S_3 as expressions of natural numbers ('0' for Zero, '$0'$' for One, etc.). Then we can introduce arithmetical symbols, for example, '$>$' for the relation Greater and '$\times$' for the function Product, respectively.[4] Let '$..x..$' indicate the matrix '$(x > 0) \bullet (x \times x \equiv x)$', which is satisfied only by the number One, hence by $0'$.] Then the following holds:

18-6. '$(\imath x)(..x..) \equiv 0'$' is true (and, moreover, L-true).

(The sentence mentioned is L-true because it holds in all state-descriptions, which differ only in the distribution of the qualitative properties. The truth of the sentence can be shown by using only the semantical rules; these include the rules determining the basic structure and the explicit and recursive definitions involved.) Hence we obtain:

[4] These and other arithmetical symbols can be introduced in a system like S_3 with the help of recursive definitions in the customary way (see, for instance, [Syntax], § 20).

18-7. The extension of '$(\imath x)(\ldots x \ldots)$' is the same as that of '$0''$'.

Hence also the following holds, because of 18-5b:

18-8. The extension of the original description '$(\imath x)(\mathrm{B}x \bullet \mathrm{C}x)$' is the same as that of '$(\imath x)(\ldots x \ldots)$'.

May we regard this statement 18-8 as a complete answer to our question? It must be admitted that it characterizes the extension of the original description only in an indirect way; this it has in common with 18-4b. In another respect, however, which is of a fundamental nature, 18-8 is different from that former answer and like those formulations which we regard as complete answers, that is, 18-5a and b and 18-2. If we receive 18-8 as an answer, then, in order to derive from it the complete and direct answer 18-5a or b, we need not make observations concerning the qualities of the positions, as in the case of the answer 18-4b; all we have to do is to carry out a certain logico-arithmetical procedure, namely, that which leads to the result 18-6. Thus there is this fundamental difference: 18-6 states an L-truth, while 18-3 states an F-truth. The following two results follow from the ones just mentioned (18-9 from 18-6, 18–10 from 18-3), according to the definitions of L- and F-equivalence (3-5b and c):

18-9. '$(\imath x)(\ldots x \ldots)$' and '$0''$' are L-equivalent.
18-10. '$(\imath x)(\mathrm{B}x \bullet \mathrm{S}x)$' and '$0''$' are F-equivalent.

It is because of the L-equivalence stated in 18-9 that we also say that 18-8 actually *gives* the extension, although indirectly. Thus it becomes clear that the difference between an answer giving the extension and one merely describing it does not simply consist in the difference between the use of a standard expression and that of a description. If a standard expression is used, the extension is certainly given; but it may also be given by a description, provided this description is L-equivalent to a standard expression, as '$(\imath x)(\ldots x \ldots)$' is, according to 18-9. If, on the other hand, a description is not L-equivalent to any standard expression, then by using it we do not give, but merely describe, the extension in question. Note that every individual expression is an expression of exactly one position and hence is equivalent to exactly one standard expression. Therefore, if an expression is F-equivalent to some standard expression, as, for example, '$(\imath x)(\mathrm{B}x \bullet \mathrm{S}x)$' is according to 18-10, then it cannot be L-equivalent to any standard expression.

The results here found will help us in constructing, in the next section, a definition for the L-determinacy of individual expressions.

§ 19. Definition of L-Determinacy of Individual Expressions

For a simple coordinate language like S_3 (§ 18), we define as L-determinate those individual expressions which are L-equivalent to standard expressions. The problem of the definition of L-determinacy for more complex coordinate languages, like the language of physics S_P, is briefly discussed. Finally, it is shown how the concept of L-determinacy can be applied also to name languages if the metalanguage is a coordinate language.

In the preceding section, we analyzed the individual expressions in the system S_3, which was chosen as an example of a coordinate language of simple structure. Analogous considerations hold for other systems in which there are individual expressions of standard form which fulfil both conditions (1) and (2), mentioned earlier. For the following definition of L-determinacy it is presupposed that S is a system for which a standard form has been determined which fulfils those conditions. This definition is suggested by the results of our discussion in the preceding section.

19-1. *Definition*. An *individual expression* in the system S is ***L-determinate*** $=_{Df}$ it is L-equivalent to an individual expression of standard form in S. (This obviously includes the standard expressions themselves.)

That this definition satisfies our previous requirement, 17-4, is seen as follows: If a given individual expression is L-equivalent to a standard expression, then those semantical rules on which this L-equivalence (in other words, the L-truth of the corresponding $\equiv$-sentence) is based suffice to give its extension, namely, the position corresponding to the standard expression. On the other hand, if a given individual expression is not L-equivalent to a standard expression, then it is, as we have seen, F-equivalent to a standard expression. Therefore, in this case the semantical rules do not suffice to give its extension; this can be given only by a factual statement.

It should be noticed that there is, in general, no effective decision procedure for the concept of L-determinacy just defined. Still less is there a general effective procedure for the evaluation of any given L-determinate individual expression, that is, for its transformation into an L-equivalent standard expression. Going back to the example of the system S_3 with arithmetical symbols (see the explanations preceding 18-6), '$(\imath x)(x \equiv 0'' \times 0'')$' can be transformed into '$0''''$' simply by calculation, that is to say, by repeated application of the recursive definitions. On the other hand, the transformation of '$(\imath x)(\ldots x \ldots)$', i.e., '$(\imath x)[(x > 0) \bullet (x \times x \equiv x)]$', into '$0'$' requires the proof of a universal arithmetical theorem, which states that every number except 1 lacks the describing property; and it is clear that there cannot be a fixed effective procedure for finding proofs of this kind. In cases like the two examples in S_3 just given, the L-determinacy is easily established by the fact that both descriptions do not contain any nonlogical constants. If, however, nonlogical constants occur, then we have, in general, no effective procedure for deciding about L-determinacy.

The basic order of the positions in a coordinate language *S* may be quite different from the simple order in S_3; but the procedure leading to a definition of L-determinacy will still be essentially the same. We first choose among the individual expressions of the system those which we wish to regard as of standard form. The choice is fundamentally a matter of convention, provided that, of the requirements stated earlier, at least the first is fulfilled. The simplicity of the forms and the possibility of recognizing the positional relations in a simple way will usually influence the choice. If the primitive constants of the language system are divided into logical and descriptive (i.e., nonlogical) constants (see § 21), then only expressions in which all constants are logical will be taken as standard form.

As an example of a system with a different basic order, let us briefly consider a coordinate language of physics S_P, leaving aside the technical details. Here the individuals are space-time points within a coordinate system chosen by convention. First, a standard form for expressions of real numbers in S_P must be chosen. Here this is a much more complicated task than in the case of natural numbers (as in S_3). The standard expressions must enable us to find the location of positions and the distance between two positions with any desired degree of precision. This means that for the representation of real numbers as systematic (e.g., decimal or dual) fractions, we must have an effective procedure for computing any required number of digits.[5] Since a space-time point is determined by three space coordinates and one time coordinate, a standard individual expression in S_P will consist of four standard real-number expressions.

A continuous coordinate language like S_P is, in certain respects, funda-

[5] This requirement can be stated in exact terms as follows. For every real number there is a unique representation in the decimal system if we exclude decimals which, from a certain place on, contain only the figure '9'. The integral part is a natural number; the fractional part corresponds to a function $f(n)$ whose value gives the nth digit after the decimal point. (For example, for $\pi = 3.1415\ldots$, $f(1) = 1$, $f(2) = 4$, $f(3) = 1$, $f(4) = 5$, etc.) If, then, a real-number expression consists of an expression of its integral part (say, in the ordinary decimal notation) and an expression for the function f corresponding to its fractional part, then this real-number expression is computable if the expression for f is computable in the sense of A. M. Turing ("On Computable Numbers", *Proc. London Math. Soc.*, Vol. XLII [1937]). Turing has shown that this concept of the computability of a function coincides with Church's lambda-definability and with the concept of general recursiveness due to Herbrand and Gödel and developed by Kleene (see Turing, "Computability and λ-Definability", *Journal of Symbolic Logic*, Vol. II [1937]).

mentally different from a discrete coordinate language like S_3. The first important difference consists in the fact that no language (with expressions of finite length) can contain expressions for all real numbers.[6] Therefore, S_P cannot contain individual expressions for all individuals, that is, space-time points—let alone individual expressions of standard form. Thus here the second of the two conditions for standard expressions cannot be fulfilled; only the first is required. Another difference is the following: There is no general effective method which would enable us to decide for any two standard individual expressions whether or not they are equivalent, that is, refer to the same position—in other words, whether or not their (four-dimensional) distance is 0. However, if two standard expressions are given, we can determine their distance in the form of a computable function. Hence, for any positive rational number δ, no matter how small it may be chosen, we can establish either that the distance is $\geqq \delta$ and hence that the positions are distinct, or that the distance is $\leqq \delta$, that is, the positions are either identical or certainly not farther apart than δ.

We cannot here go any further into the technical details of the problem of L-determinacy for the individual expressions in S_P. The problems which ought to be investigated are the following. It is clear that not all individual expressions in S_P can be equivalent to standard expressions. The question should be examined as to whether the standard form can be chosen in such a manner that at least all those individual expressions which do not contain nonlogical constants are equivalent (and hence L-equivalent) to standard expressions. If so, L-determinacy can be defined for S_P as in 19-1. Otherwise, a more complicated definition will perhaps be necessary; but it will, in any case, be such that L-equivalence to a standard expression is a sufficient, though perhaps not a necessary, condition for L-determinacy.

So far we have applied the concept of L-determinate individual expressions only to coordinate languages. Now let us consider name languages, as, for example, S_1. In a language of this kind we have no individual expressions which exhibit their positional relations directly by their form. We may have individual expressions in the form of descriptions using qualitative describing properties; furthermore, there may be individual constants which are either primitive or perhaps introduced by definitions as abbreviations of descriptions. However, even a primitive individual constant in a name language *S* may, under certain conditions, be L-determinate if the metalanguage *M* is a coordinate language. For every primitive individual constant in *S* there is a rule of designation in *M*

[6] See [Syntax], § 60d.

which tells us which individual is meant by the expression. This rule refers to the individual by an individual expression in *M*. Now if *M* is a coordinate language and the individual expression used in the rule is L-determinate in *M* in the sense earlier explained for coordinate languages, then we may likewise regard the individual constant in *S* as L-determinate. This extended use of the term 'L-determinate' seems natural, since it satisfies our earlier requirement 17-4: The semantical rules give the extension of the constant, that is, the location of the position to which the constant refers. This may be illustrated by the following example. Suppose the expressions 'o', 'o′', 'o″', etc., occur, not in the object language *S*, which is supposed to be a name language with individual constants 'a', 'b', etc., but in *M*, and that they refer, as explained earlier for S_3, to the positions in a discrete linear order. Suppose, further, that the following two rules are among the semantical rules of *S* formulated in *M;* they are rules of designation for the primitive constants 'a' and 'b':

19-2. a. 'a' designates the position o″.

b. 'b' designates either the one position which is both blue and cold, or the position o if no or several positions are blue and cold.

We would in this case construct the definition of L-determinacy in such a way that 'a' will be called L-determinate but 'b' not. (We omit here the actual construction.) These results will then be in agreement with the requirement 17-4. We see from rule 19-2a that the extension of 'a' is the third position. On the other hand, the semantical rules do not give the extension of 'b' but merely describe it (in rule 19-2b); it can be given only by the addition of a factual statement to the rules. Thus the first part of the factual statement 18-1, together with the rule 19-2b, tells us that the extension of 'b' is the second position (o′).

§ 20. L-Determinacy of Predicators

A predicator (in a coordinate language like S_3) is said to be L-determinate if every full sentence of it with individual expressions of standard form is L-determinate. This holds if the intension of the predicator is a positional or mathematical, rather than a qualitative, property. The analogous definition for functors is briefly indicated.

The concept of the extension of a predicator, especially if we consider predicators of degree one, seems entirely clear and unproblematic. For example, the extension of the predicator 'H' in the system S_1 is the class Human because its intension is the property Human. We began the explanation of the method of extension and intension with the customary and apparently clear and simple distinction between classes and properties (§ 4). We took this distinction as a model and framed the distinctions between the extension and the intension of sentences and of individual expressions in analogy to it (§§ 6 and 9). A closer inspection shows, however, that a serious difficulty is involved even in the concept of the extension of a predicator. We could leave this difficulty aside in our earlier discussions, but for our present purpose we have to face it and try to overcome it. In order to find an adequate definition for L-determinacy of predicators we have to make clear the means by which a class can be given. We shall see presently that this problem cannot be solved without first solving the problem of the way in which the extension of an individual expression can be given. This was our reason for first discussing individual expressions in the two preceding sections.

Suppose we ask somebody for information about the extension of the membership of Club C; that means that we want to learn who is a member of C and who is not. The answer 'the extension is the class of the members of C' is, although true, entirely trivial and hence would not satisfy us. Nor would an answer like 'the class of those boys in this town who either are between fifteen and sixteen years old or have red hair'. Although this answer is not trivial, it still does not *give* the extension but merely describes it with the help of another complex property which happens to have the same extension. What we want is not an indirect characterization of the membership by an intension but a membership list. Would every kind of membership list satisfy us? We see easily that some kinds would not. Thus the problem arises: What kind of membership list does actually give the extension? Suppose that we are given a statement which lists all the members of the club but does so by formulations like these: 'the eldest son of Mr. Jones', 'the boy friend of Mary', etc. We should again reject this statement, although it enumerates all members, because it does so by descriptions. Thus we see that a certain class is not merely described but actually given by a statement if this statement (1) refers to each of the members of the class and (2) does so by the use of individual expressions, which, in turn, do not merely describe but give the individuals—in other words, by the use of L-determinate individual expressions. This shows that the concept of L-determinacy of predicators presupposes the concept of L-determinacy of individual expressions.

We presuppose for the following discussions that *S* is a coordinate language of a simple structure similar to S_3, as explained in the beginning of § 19; that a standard form of individual expressions has been defined for *S;* and that L-determinacy of individual expressions in *S* is defined by our previous definition (19-1).

The condition formulated above for a statement giving a class is sufficient but not necessary. The statement need not give an enumeration of all members of the class; if this were necessary, then only finite classes could be given. It is sufficient and also necessary that the statement logically imply the truth of all those true singular sentences in *S* which say of an individual that it is or that it is not a member of the class, where the individual expressions occurring are L-determinate.

It would even be sufficient to require this merely for all the individual expressions of standard form in *S;* it is easily seen that it also holds, then, for all L-determinate individual expressions because they are L-equivalent to standard expressions, according to the definition 19-1.

In order to give examples let us go back to the coordinate language S_3 with '0', '0′', '0″', etc., as standard expressions. Suppose that the statement 'the positions 0 and 0‴ and no others are blue' is true. Then it gives the extension of the predicator 'B', because from this statement, together with the semantical rules, we can infer that 'B(0)' and 'B(0‴)' are true, while all other full sentences of 'B' with a standard expression are false. Let us introduce into S_3 the customary notation '{. . , . . , . .}' for a finite class indicated by an enumeration of its members; the definition can be written with the help of a lambda-operator as follows:

20-1. *Abbreviation.* '$\{x_1, x_2, \ldots x_n\}$' for '$(\lambda y)[(y \equiv x_1) \vee (y \equiv x_2) \vee \ldots \vee (y \equiv x_n)]$'.

Then the extension of 'B' in the above example can be given also by this statement: 'the extension of 'B' is the same as that of '{0, 0‴}'.

These considerations suggest the following definition for L-determinacy of predicators in a system *S* (of the kind indicated above). It presupposes the definition of L-determinacy for sentences (2-2d).

20-2. *Definition.* A *predicator* in *S* is ***L-determinate*** $=_{\mathrm{Df}}$ every full sentence of it with individual expressions of standard form is L-determinate.

We see easily that this definition fulfils our earlier requirement 17-4; the concept defined applies if and only if the semantical rules alone, without any factual knowledge, suffice to give the extension of the predicator in the sense explained above, because a sentence is L-determinate if and only if the semantical rules suffice to determine its truth-value (convention 2-1).

We see that any predicator in S_3 of the form '$(\lambda x)(\ldots x \ldots)$', where any molecular combination of '$\equiv$'-matrices with 'x' and standard expressions stands in the place of '$\ldots x \ldots$', is L-determinate. Therefore, '{0, 0‴}' is L-determinate, and likewise any other predicator of the form '{. . .}'

where all individual expressions occurring are of standard form. Let us define in the customary way the signs 'Λ' and 'V' of the null class and the universal class, respectively, or, more exactly, of the L-empty property and the L-universal property, respectively:

20-3. *Abbreviations.*
a. 'Λ' for '$(\lambda x)[\sim(\mathrm{x} \equiv \mathrm{x})]$'.
b. 'V' for '$(\lambda x)[\mathrm{x} \equiv \mathrm{x}]$'.

We see immediately that the two predicators here defined are L-determinate, because all full sentences of 'Λ' are L-false and all of 'V' are L-true. But there are other, more complicated predicators which likewise are L-determinate, among them all lambda-expressions with any purely arithmetic conditions. Take, as an example, the predicator '(λx)[Prime (x)]', where 'Prime' is defined so that it holds for all prime numbers (that means, for all positions with a prime coordinate).[7] This example shows that even a predicator whose extension is infinite and therefore cannot be given by an enumeration may be L-determinate. This is the case if the intension is of a mathematical, rather than of an empirical, nature; in other words, if the intension is a positional, rather than a qualitative, property. That, for instance, the position $0'''$, corresponding to the number Three, belongs to the extension of 'Prime' is found by a purely logico-mathematical procedure, that is, a procedure based upon the semantical rules and not involving the qualitative properties of that or any other position. On the other hand, for establishing that the position $0'''$ belongs to the extension of 'B', we need not only the semantical rules but, in addition, an observation yielding the result that this position has the color Blue.

Here, again, for the concept of L-determinate predicators there is no effective method of decision, since there is none for the concept of L-determinate sentences on which it is based. For example, let x be called a Fermat exponent if $x > 2$ and if there are positive integers u, v, and w such that $u^x + v^x = w^x$. A predicator for this property, say 'Fer', can easily be defined in S_3. 'Fer' is an L-determinate predicator because every full sentence of it with a standard individual expression is an L-determinate sentence. For most of these sentences it is at present unknown whether they are true or false, and there is no decision method for determining their truth-value. Nevertheless, they are L-determinate, because their truth-values are independent of colors or any other qualitative properties of the corresponding positions. For the number Three and some others it is known that they are not Fermat exponents. This has been

[7] Arithmetical concepts of this kind can be defined in a language similar to S_3 with the help of recursive definitions (see, for example, [Syntax], § 20).

shown by a mathematical proof; thus the result is independent of the qualitative properties of the positions. Therefore, the sentence '$\sim$Fer(o‴)' holds in every state-description and hence is L-true in S_3.

It may be remarked incidentally that a definition of L-determinacy for functors and compound functor expressions can be given which is quite analogous to that for predicators (20-2). Here it would likewise be required that every full sentence in which the argument expressions and the value expression are of standard form be L-determinate. Thus, all signs or expressions for arithmetical functions are L-determinate. For example, the functor '+' in S_3 is L-determinate because every full sentence with standard expressions is L-determinate; for instance, 'o′ + o″ ≡ o‴' is L-true. On the other hand, a functor for a physical magnitude, for example, temperature (say, in the language of physics, S_P) is not L-determinate, because a sentence saying that the temperature at a certain space-time point has a certain value is not L-determinate.

§ 21. Logical and Descriptive Signs

We make use in this section of the customary distinction between logical and descriptive (nonlogical) signs. For the system S_3 (restricted to primitive signs) the classification is simple: the primitive predicates are descriptive, all other signs are logical. If a designator in S_3 contains only logical signs, then it is L-determinate. A designator in S_3 is L-determinate if and only if it is L-equivalent to a designator containing only logical signs. This could be taken as an alternative way of defining L-determinacy.

In this section we make the customary distinction between logical and descriptive, i.e., nonlogical signs.[8] With its help we shall then make a corresponding distinction for expressions, which is especially important for designators. Then we shall investigate the relation between this distinction and the distinction between L-determinate and L-indeterminate designators. The concepts of logical and descriptive signs will seldom be used in the rest of the book.

We shall define the concepts mentioned for two example systems, one a coordinate language and the other a name language. As coordinate language we take the system S_3 discussed in the preceding sections; it contains 'o', 'o′', etc., as individual expressions of standard form. As name language we take a system S_1' which is like our system S_1 with this exception: We suppose that the individual constants in it, say 'a*', 'a', 'b', etc., are interpreted by the semantical rules of S_1' as referring, not to things, as in S_1 (see rule 1-1), but to positions in an ordered domain (as, for ex-

[8] For more detailed explanations see [I], § 13.

ample, in rule 19-2a). Therefore, these constants are L-determinate, as explained earlier (at the end of § 19). Both systems are here supposed to contain only primitive signs, not defined signs. The predicates in both systems are supposed to be interpreted by the semantical rules as designating qualitative properties or relations like Blue, Cold, Colder, and the like (as explained for S_3 in § 18).

The distinction between logical and descriptive signs of the systems S_3 and S'_1 is made in the following way by enumeration of particular signs and kinds of signs.

21-1. The following *signs* are regarded as *logical:*

a. The individual variables.

b. The connectives; the operator signs '∃', '℩', 'λ'; the parentheses.

c. In S'_1, the individual constants; in S_3, '0' and '′'.

21-2. The predicates are regarded as *descriptive signs.*

The corresponding distinction for expressions is now defined in 21-3; to be descriptive is taken, so to speak, as a dominant property; to be logical as a recessive property.

21-3. *Definitions.*

a. An *expression* is *logical* $=_{Df}$ it contains only logical signs.

b. An *expression* is *descriptive* $=_{Df}$ it contains at least one descriptive sign.

Thus the standard expressions '0', '0′′', etc., in S_3 are regarded as logical. This seems justified because they refer here not to things but to positions in a basic, presupposed order. We may even interpret them as referring to pure numbers. In a word translation of 'B(0′′′)' the expression '0′′′' corresponds in this interpretation to the italicized part in 'the position correlated to *the number Three* is blue', while the predicate 'B' corresponds to the whole nonitalicized part of this sentence.[9] This interpretation is just as adequate as the ordinary interpretation by '*the position correlated to the number Three* is blue'. We might even say that these are merely two different formulations for the same interpretation, since the translation of the whole sentence is the same in both cases, and hence the truth-condition of the sentence remains likewise the same.

In addition to the individual expressions of standard form in S_3 (e.g., '0′′′') and in S'_1 (here we take the individual constants as standard form), both systems contain individual descriptions.

The following results concern the system S_3. They hold likewise for S'_1,

[9] This interpretation has, furthermore, the advantage that a sentence which says that the universe of individuals is infinite is not factual but L-true. Thus the difficulty usually connected with the so-called Axiom of Infinity is here avoided (see [Syntax], p. 141).

provided that the basic order of its universe of individuals is either the same as in S_3 or has a similar simple structure and provided that the rules of designation formulated in *M* for the individual constants in S'_1 use only individual expressions of standard form; this standard form in *M* may, for example, be the same as in S_3.

21-4. Every sentence in S_3 which contains only logical signs is either L-true or L-false; and there is an effective decision method for determining which of the two is the case.

21-5. Every (closed) description in S_3 is L-determinate; and there is an effective procedure for transforming it into an individual expression of standard form.

21-6. Every closed lambda-expression in S_3 is L-determinate; and there is a decision method for any full sentence of the lambda-expression with any individual expression of standard form.

The proofs of these theorems and the decision methods mentioned cannot be given here, but they are rather simple.[10] They are based on the following circumstances: (1) since no predicates occur, the ultimate components are $\equiv$-matrices; (2) an $\equiv$-sentence with two standard expressions is L-true if the two standard expressions are alike, and otherwise it is L-false.

The three results can be combined into one as follows:

21-7. Every designator in S_3 which contains only logical signs is L-determinate.

There are, however, also L-determinate designators which contain descriptive signs. For example, '$\mathrm{P}(0) \vee \sim \mathrm{P}(0)$' is L-true; '$(\lambda x)(\mathrm{P}x \vee \sim \mathrm{P}x)$' is L-universal, and hence L-equivalent to 'V' (20-3b); and '$(\imath x)(\mathrm{P}x \vee \sim \mathrm{P}x)$' is L-equivalent to '0'; thus these three designators are all L-determinate.

It follows from 21-7 that any designator L-equivalent to one containing only logical signs is likewise L-determinate. Now it can be shown that the converse of this holds too. (i) If a sentence is L-determinate, then it is either L-true or L-false; therefore, it is L-equivalent either to '$0 \equiv 0$' or to the negation of this sentence. (ii) If a description is L-determinate, it is L-equivalent to a standard expression, according to the definition 19-1. (iii) It can be shown that, if a closed lambda-expression in S_3 is L-determinate, either its extension or the complement of its extension is finite; therefore, the lambda-expression is L-equivalent to one of the form '$(\lambda x)(\ldots x \ldots)$', whose scope is constructed with the help of connectives

[10] For further details see [Modalities], §§ 11 and 12, especially T12-2f.

out of ≡-matrices with 'x' and standard expressions. Thus the following holds:

21-8. A designator in S_3 is L-determinate if and only if it is L-equivalent to one containing only logical signs.

For S_3 and similar systems, L-determinacy for designators could be generally defined by the sufficient and necessary condition stated in 21-8. This alternative method presupposes only the concepts of logical signs (21-1) and of L-equivalence of designators (3-5b), hence of L-truth of sentences (2-2); it would replace the three separate definitions of L-determinacy for sentences, individual expressions, and predicators earlier given (2-3d, 19-1, 20-2).

Now we can easily see that if two designators in S_3 which contain only logical signs are equivalent, then they are L-equivalent. Since they are equivalent, the ≡-sentence containing them as components is true (3-5a) and therefore L-true, according to 21-4; hence they are L-equivalent (3-5b). From this result the following more general theorem can be derived with the help of 21-8 and the transitivity of equivalence and L-equivalence:

21-9. If two L-determinate designators in S_3 are equivalent, then they are L-equivalent.

§ 22. L-Determinate Intensions

If a designator is L-determinate, then all designators L-equivalent to it are likewise L-determinate. We shall say of the common intension of these designators that it is an *L-determinate intension*. For any extension, there are, in general, many corresponding intensions; but there is among them exactly one L-determinate intension.

The results which will be stated here can be proved in an exact way for the system S_3. But it can be shown in an informal way that they hold likewise for any system S, provided the concepts of L-truth and L-determinacy are defined for S in such a manner that our requirements for these two concepts (2-1 and 17-4, respectively) are fulfilled. In the following discussion it is presupposed that these requirements are fulfilled.

22-1. If two L-determinate designators in S are equivalent, then they are L-equivalent.

Applied to S_3, this is the same as 21-9, which was proved with the help of the distinction between logical and descriptive signs. The general theorem for a system S can be seen to hold in the following way, which does not presuppose such a distinction. Since the two designators are equivalent, they have the same extension (5-1). Since they are L-determinate, the semantical rules suffice for establishing that both have this same extension (17-4) and hence that they are equivalent (5-1) and hence that their ≡-sentence is true (3-5a); therefore, this ≡-sentence is L-true (2-1); hence the two designators are L-equivalent (3-5b).

22-2. If a designator in *S* is L-equivalent to an L-determinate designator, then it is itself L-determinate.

For S_3, this follows from 21-8 because of the transitivity of L-equivalence. That it holds generally for *S* is seen as follows: If the condition in 22-2 is fulfilled, the semantical rules suffice for establishing the extension of the second designator and the identity of extension for the two designators, and thereby the extension of the first designator.

Suppose an L-determinate designator in *S* is given. It possesses a certain intension. Any other designator having this same intension is L-equivalent to the first and hence likewise L-determinate, according to 22-2. Let us call an intension of this kind an ***L-determinate intension.*** Thus, roughly speaking, an L-determinate intension is such that it conveys to us its extension. For every extension, there are, in general, many corresponding intensions; but among them there is exactly one L-determinate intension, which may, in a way, be regarded as the representative of this extension (not, of course, in the sense in which a designator may be said to represent, or refer to, its extension). This one-one correlation between extensions and L-determinate intensions will become clearer with some examples.

For *sentences*, there are only two extensions, the two truth-values, Truth and Falsity. There are many L-determinate sentences whose extension is the truth-value Truth, namely, all the L-true sentences, e.g., 'Pa ∨ ∼Pa' (in S_1). Since they are L-equivalent to each other, they have the same intension, namely, the L-true or necessary proposition. Thus this proposition is the one L-determinate intension corresponding to the extension Truth. Analogously, the L-false or impossible proposition is the L-determinate intension which corresponds to the extension Falsity. For *predicators*, there are infinitely many extensions, namely, classes of individuals. If, as in S_1 and S_3, the number of individuals is denumerably infinite, the number of classes of individuals is nondenumerable; since the number of (finite) expressions in any language system *S* is, at most, denumerable, not all classes of individuals can be extensions of predicators in *S*. For an extension referred to by a predicator in *S* there is not necessarily always a corresponding L-determinate intension expressed by a predicator in S, because not every predicator has an equivalent L-determinate predicator. Whether a certain L-determinate intension is or is not expressed by a predicator in S depends on the means of expression in S. The L-determinate intension corresponding to the null class of individuals is the L-empty property; in S_1 and S_3 this intention is expressed, for example, by '$(\lambda x)[\sim(x \equiv x)]$'. The L-determinate intension corresponding to the universal class is the L-universal property, expressed by '$(\lambda x)[x \equiv x]$'. The L-determinate intension corresponding to the class whose only members are the positions o, o″, and o‴ is the property of being one of these three positions, which is expressed in S_3 by '$(\lambda x)[(x \equiv \text{o}) \vee (x \equiv \text{o}'') \vee (x \equiv \text{o}''')]$'. On the other hand, suppose that the primitive signs of S_3, mentioned earlier, are the only signs in S_3 and that S_3' is constructed from S_3 by the addition of some recursively defined functors and predicators, among them the predicator 'Prime' for the property Prime Number. Suppose, further, that all prime number positions, and only these, happen to be blue. Then the extension of 'B' is the class of prime number positions, and the corresponding L-determinate intension is the property of being a prime number position. This intension is expressed in S_3' by the L-determinate predicator 'Prime'; but in S_3 it is not expressed by any predicator.

The extensions of *individual expressions* are the individuals, which in S_3 are the positions. For example, the extension of the description '$(\iota x)(\text{B}x \bullet \text{C}x)$' in our earlier example is the second position (i.e., the position next to the initial position, 18-5a). Therefore, the corresponding L-determinate intension is the individual concept The Second Position, which is expressed in S_3 by the L-determinate individual expression 'o′'. Generally speaking, for every individual in S_3 there is one L-determinate intension, namely, the individual concept of that position; this intension is expressed in S_3 by at least one L-determinate individual expression, for instance, by the standard expression ('o', 'o′', etc.).

§ 23. Reduction of Extensions to Intensions

The one-one correlation between extensions and L-determinate intensions suggests the identification of extensions with the corresponding L-determinate intensions. According to this method, which is discussed in this section but will not be used in the remainder of the book, a class is construed as a positional property. This leads to explicit definitions of classes, in distinction to the contextual definitions used by Whitehead and Russell.

The method of extension and intension introduced in the first chapter assigns to every designator an extension and an intension. Thus our semantical analysis of the designators seems to assume two kinds of entities—extension and intensions. It has been mentioned earlier that this assumption is not actually made, that, in fact, we merely use two forms of speech which can ultimately be reduced to one. There are several possibilities for this reduction; they fall chiefly into three kinds: (i) the extensions are reduced to intensions; (ii) the intensions are reduced to extensions; (iii) both extensions and intensions are reduced to entities, which are, so to speak, neutral. We shall later explain several methods of the first kind. The chief requirement that such a method must fulfil is obviously this: two different but equivalent intensions must determine the same extension. The methods of this kind to be explained later (§ 33, methods (2) and (3)) give, not an explicit definition, but only a contextual one. That is to say, a phrase like 'the class Blue' is not itself translated into a phrase in terms of properties; instead, a rule is given for transforming any sentence containing the phrase 'the class Blue' into a sentence referring only to properties.

Now the introduction of the concept of L-determinate intension (in the preceding section) makes it possible to define extensions in terms of intensions. This method requires that the universe of individuals in question exhibit a basic order so that the concept of L-determinacy may be applied. It is not required that the object language be a coordinate language; the basic order need not be exhibited by the individual expressions of the object language; it is sufficient that it be expressible in the metalanguage. We suppose for the following definitions, as we did in the preceding section, that the concepts of L-truth and L-determinacy are defined for the system S in such a manner that our requirements for these two concepts (2-1 and 17-4) are fulfilled. The advantage of the method to be applied here is that it supplies explicit definitions. It is based on the following three results, which we found earlier: (i) to every intension there corresponds exactly one L-determinate intension; (ii) the L-determinate intensions corresponding to any two intensions which are equivalent and hence have the same extension are identical; (iii) therefore, there is a one-one correlation between extensions and L-determinate intensions.

The method to be proposed consists simply in identifying extensions with the corresponding L-determinate intensions.

23-1. *Definition.* The *extension of a designator* in S $=_{Df}$ the one L-determinate intension which is equivalent to the intension of the designator.

The concept of the equivalence of intensions used in this definition was introduced (definition 5-3) with the help of the concept of the equivalence of designators; the latter concept was defined (3-5a) by the truth of an $\equiv$-sentence and hence does not presuppose the concept of extension.

Our principal requirement for extensions was that they be identical for equivalent designators (5-1). This requirement is fulfilled by the present definition 23-1 (see (ii) above).

Although we have usually spoken of intensions only as intensions of designators, occasionally reference was made to intensions independent of the question of whether or not they were expressed by designators in the system under discussion. Therefore, it may be useful to have the following definition for the extension of (or, corresponding to, determined by) an intension; here no reference is made to designators.

23-2. *Definition.* The *extension of a given intension* $=_{Df}$ the one L-determinate intension which is equivalent to the given intension.

Let us apply these definitions to the examples in the system S_3 given in the preceding section. Let us begin with *predicators*, because in this case the concept of extension, that is, of class, is more familiar than in the other cases. Classes are now identified with L-determinate properties, that is, positional properties. Let us assume, for example, that the positions o, o″, o‴, and no others, are blue. On the basis of this assumption, the extension of the predicator 'B' in S_3 is, according to the definition 23-1, the intension of '$(\lambda x)[(x \equiv o) \vee (x \equiv o'') \vee (x \equiv o''')]$', that is, the property of a position of being either o or o″ or o‴. And we say likewise, according to the definition 23-2, that the extension of the property Blue is the positional property just mentioned. However, it should be noted that these two results are factual statements based on the factual assumption mentioned. Our definitions do by no means say that the phrases 'the extension of 'B' ' and 'the extension of the property Blue', to which we may add the third synonymous phrase, 'the class Blue', mean the same as 'the property of being either o or o″ or o‴'. The latter phrase is merely equivalent to each of the three former phrases. What the definition 23-1 actually says is that the phrase 'the extension of 'B' ' means the same as 'the L-determinate intension which is equivalent to the intension of 'B' '—in other words, 'the positional property which is equivalent to the (qualitative) property Blue'. It is a matter of fact, not of logic, that the positional property which is equivalent to the property Blue is the property of being either o or o″ or o‴.

Let us assume, further, that no position is both blue and cold. Then the extension of '$B \bullet C$' in S_3 is the null class; this is now identified with the L-empty property, which is expressed in S_3 by the predicator

'$(\lambda x)[\sim(x \equiv x)]$'. Suppose that all prime number positions, and no others, are blue. Then the extension of 'B' is the class of prime number positions. This class is now identified with the property Prime Number Position.

It may perhaps at first seem somewhat strange to regard classes not as distinct entities corresponding somehow to properties but as properties of a special kind. But a consideration of the examples given will remove or mitigate the feeling of strangeness. For example, it might not seem very unnatural to regard the intension of '$(\lambda x)[(x = \mathrm{o}) \vee (x = \mathrm{o}'') \vee (x \equiv \mathrm{o}''')]$' as a class when we consider the fact that this intension, in contrast to L-indeterminate intensions, provides by itself an answer to the question as to the individuals to which it applies and those to which it does not.

Now we are going to apply our definitions to *sentences*. If we approach the matter naïvely, without careful analysis as to the nature of the entities, we might perhaps be inclined to say that we know, at least roughly, what we mean by the extension of a predicator (of degree one), that is, a class. However, if it is said that the extension of a sentence is a truth-value, it is not at all clear what entities should be regarded as truth-values. In our earlier discussion (in § 6), we left aside the difficulty here involved; but now let us examine it and try to solve it. We consider here languages which speak about extra-linguistic individuals, either physical things with physical properties, as in S_1, or positions, as in S_3, with physical properties (e.g., 'the second position is cold'). Both the intensions and the extensions of predicators are clearly extra-linguistic entities; both properties of individuals and classes of individuals (no matter whether regarded in the customary way or, according to the method here proposed, as properties of a special kind) have to do with the individuals, not with expressions in the language. The same holds for extensions and intensions of individual expressions; both individuals and individual concepts, whatever their specific nature may be, are certainly extra-linguistic entities. Therefore, it seems natural to expect, by analogy, that intensions and extensions of designators of all kinds are extra-linguistic entities. This holds also for the intensions of sentences, the propositions. But what about their extensions? What kind of entities are the truth-values which we take as the extensions of sentences? We might perhaps be inclined to answer that the truth-values are truth and falsity and that these two terms are to be understood in their semantical sense. However, truth in the semantical sense is a certain property of sentences, hence a linguistic entity. [This does not imply that truth is a merely linguistic matter; truth is dependent upon extra-linguistic facts; therefore, its definition must refer to extra-linguistic entities. However, we are here not concerned with the question of the entities to which the definition refers, but rather with the question of the kind (logical type) of entity to which the concept of truth belongs. And here the answer is: It is a property of sentences.] Therefore, truth and falsity fall outside the domain to which all other intensions and extensions belong. Now there is nothing in the situation that compels us to take (semantical) truth and falsity as the extensions of sentences. All that is required is that the extension of all true sentences be the same entity and that the extension of all false sentences be the same entity but something different from the first. There are obviously many different possibilities of choosing in a not too arbitrary manner two extra-linguistic entities such that the one is connected in a simple way with all true sentences and the other with all false sentences. What type of nonlinguistic entities should we choose? It seems most natural to choose either two properties of propositions or two propositions. Let us consider some possibilities of these two kinds. The most natural properties of propositions to be considered would obviously be truth and falsity of propositions. [In distinction to truth or falsity of sentences, these two concepts are not semantical but independent of language.[11] Their relation to the semantical concepts of truth and falsity is the same as the relation of the equivalence of intensions to the equivalence of designators; see the definition 5-3 and the explanations preceding it, including the footnote. They are singulary, truth-functional connections.[12]] It would be simpler to take two propositions. We might, for example, take, on the one hand, the proposition p_T expressed by the class of all true sentences in *S*, and, on the other hand, the negation of p_T. [In systems like S_1 and S_3, where we have state-descriptions (§ 2), the proposition p_T is expressed in a simpler way by the one true state-description.] This device might perhaps appeal to those philosophers who regard truth as involving in some sense the whole universe.[13] While this method takes two factual (contingent) propositions as extensions, our own method (23-1) takes the two L-determinate propositions. Here the extension of any true sentence is the L-true (necessary) proposi-

[11] In the terminology of [I], they are absolute concepts; for their definitions, see [I], D17-1 and D17-2.

[12] 'True' in this sense is a connective with the characteristic TF and hence is redundant (e.g., '(the proposition) that Scott is human is true' and 'Scott is human' are L-equivalent sentences in M); 'False' has the characteristic FT and hence is a sign of negation (compare [II], § 10).

[13] Lewis ([Meaning], p. 242) maintains a similar conception. The denotation or extension of a proposition "is not that limited state of affairs which the proposition refers to, but the kind of *total* state of affairs we call a world. . . . All *true* propositions have the same extension, namely, this actual world; and all *false* propositions have the same extension, namely, zero-extension."

tion; and the extension of any false sentence is the L-false (impossible) proposition. Here, likewise, we probably feel, at first, some reluctance to regard propositions as truth-values or extensions. However, the connection between the two L-determinate propositions and what we usually regard as the truth-values is so close and natural that it is perhaps not too artificial to take these propositions as extensions of sentences.

Now let us apply the new method to *individual expressions*. Let us again assume that only the second position o′ in S_3 is both blue and cold. We said earlier that, on the basis of this assumption, the extension of '$(\imath x)(Bx \bullet Cx)$' is the second position. We say now, instead, that the extension of this description is the individual concept The Second Position. In a sense this may be regarded as merely a change in formulation. We may even use the same formulation as before, by saying: "The extension of the description is o′". The change appears only when we add to 'o′' a specifying noun. But this addition serves merely for greater clarity. The new method does not lead to the result 'the extension is not the individual (or position) o′'. The situation is, rather, this: the new method in its primary formulation does not use the terms 'individual', 'class', 'truth-value' at all; thus 'o′' and 'the individual concept o′' are synonymous. In a secondary formulation those terms might be reintroduced under the new method, in analogy to the introduction of 'extension' by 23-1 and 23-2. But then again a combination of any of these three terms with an L-determinate designator is synonymous with the designator alone. Thus, for example, on this method the phrases 'the individual (or position) o′', 'o′', and 'the individual concept o′' all mean the same. Likewise, if 'Λ' is used in M, the phrases 'the class Λ' (or 'the null class'), 'Λ', and 'the property Λ' all mean the same.

I will not decide here the question of whether the method of taking L-determinate intensions as extensions is or is not natural. It may suffice to have shown that this method meets the formal requirements of a solution to the problem of extensions. For the further discussions in this book, this method will not be presupposed; most of the discussions will be independent of any particular specification of the nature of the entities chosen as extensions, beyond the general requirement that equivalent designators have the same extension (5-1).

CHAPTER III

THE METHOD OF THE NAME-RELATION

The method of the name-relation is an alternative method of semantical analysis, more customary than the method of extension and intension. It consists in regarding expressions as names of (concrete or abstract) entities in accordance with the following principles (§ 24): (1) every name has exactly one nominatum (i.e., entity named by it); (2) any sentence speaks about the nominata of the names occurring in it; (3) if a name occurring in a true sentence is replaced by another name with the same nominatum, the sentence remains true. An examination of the method shows that its basic concept involves an essential ambiguity (§ 25) and that it leads to an unnecessary duplication of expressions in the object language (§§ 26, 27). The most serious disadvantage of the method consists in the fact that the third of the principles mentioned, although it seems quite plausible, leads in certain cases to a contradiction if applied without restriction; we call this contradiction the antinomy of the name-relation (§ 31). It is not difficult to eliminate the contradiction; various ways have been proposed by logicians, but all of them have certain drawbacks. The method of Frege is discussed in detail (§§ 28–30). Its main feature is the distinction between the nominatum and the sense of an expression. In many cases these are the same as what we call the extension and the intension, respectively. However, in contradistinction to these latter concepts, the nominatum and the sense of an expression vary with the context in which the expression occurs. It is found that Frege's method, if applied consistently, leads to an infinity of new entities and new expressions as names for them and thus results in a very complicated structure of the object-language. This holds still more for the variant of Frege's method proposed by Church. Russell and Quine avoid the antinomy by not regarding as names certain expressions (although these expressions are, in our method, L-equivalent to other expressions, which they do regard as names); thus they require an unnecessary restriction of the field of application of semantical meaning analysis (§ 32). The fact that all forms of the method of the name-relation lead to complications or restrictions makes it appear doubtful whether this method is a suitable method of semantical analysis.

§ 24. The Name-Relation

The customary method of meaning analysis regards an expression as a *name* for a (concrete or abstract) entity, which we call its ***nominatum***. The method, as customarily used, is based on three principles, usually implicit: the principles of univocality, of subject matter, and of interchangeability.

In chapter i the concepts of equivalence and L-equivalence were introduced and discussed, together with the derivative concepts of the extension and the intension of an expression. These concepts have been proposed as tools for a semantical analysis of meaning. With our method of extension and intension we shall now contrast that method of analysis which seems to be accepted by many, probably by most, logicians; it is characterized by using as basic concept the name-relation. In the present chapter the assumptions underlying this method of the name-relation will be made explicit, and the consequences of its use investigated. It will be shown that the method leads to certain difficulties, one of which will be called the antinomy of the name-relation. Some of these difficulties have been recognized by several logicians, and various ways have been proposed to avoid them, thus leading to different forms of the method of the name-relation. An examination of these forms will show that each of them has serious disadvantages, e.g., an intrinsic ambiguity in the terms used, an unnecessary multiplication of the entities leading to a complicated language structure, or unnecessary restrictions in the construction of languages. It will be seen that the method of extension and intension is free of the shortcomings which the customary method of the name-relation shows, at least in its known forms.

The name-relation is customarily conceived as holding between an expression in a language and a concrete or abstract entity (object), of which that expression is a name. Thus this relation is, in our terminology, a semantical relation. Various phrases are used to express this relation, e.g., '*x* is a ***name*** for *y*', '*x* denotes[1] *y*', '*x* designates *y*', '*x* is a designation for *y*', '*x* signifies *y*', etc. In this book I shall sometimes also use, besides '*x* is a name of *y*', '*x* names *y*'; this shortened form will not lead to any ambiguity, since its customary meaning ('a person names an entity') will hardly occur here. It is often convenient to have a short term for the converse relation; I shall often say, instead of 'the entity named by (the expression) *x*', 'the ***nominatum*** of *x*'; I shall use this term also in formulating the conceptions of other authors who do not use it.

Logicians seem to differ widely with respect to the question of the kinds of expressions which may be regarded as names. Nearly all will

[1] The phrase '*x denotes y*' is often used in a quite different sense, namely, in the case where *x* is a predicator for a certain property (e.g., the word 'human') and *y* is an entity having that property (e.g., the man Walter Scott). This semantical relation is of a rather special kind, since it is applicable not to designators in general but only to predicators and, moreover, only to predicators of degree one, unless one is willing to regard a sequence of entities as the entity denoted. As a term for this relation, perhaps '*x* applies to *y*' and the corresponding noun 'application' might also be considered. In any case, the word 'denotes' is at present used by many logicians in the sense of the name-relation (see Church, [Dictionary], p. 76). Russell ([Denoting]) has used the word in this sense both for the formulation of his own conception (he uses, for instance, the term 'denoting phrases' for descriptions and similar expressions) and as a translation for Frege's term 'bezeichnet' (see below, § 28, n. 21). Church likewise uses this word for the formulation of his conception, which is based on Frege's. Following Russell and Church, I used the word 'denotes' for the name-relation in the first version of this book. However, in view of the ambiguity just described, I now prefer to avoid it.

include words like 'Napoleon' or 'Chicago'; perhaps a majority also words like 'green' (or 'greenness'), 'house', and 'seven'; many also (declarative) sentences. Let us disregard at present these differences in the domain of application of the relation and look, rather, at the way in which it is applied. It seems to me that many logicians use the name-relation for semantical discussions, that is, for speaking about expressions and their meanings, in such a way that the following three principles are fulfilled. If an author fulfils these conditions, then we shall say that he uses the ***method of the name-relation,*** irrespective of the terms he may use for the relation. Sometimes an author may state the principles explicitly; more often we shall have to infer from the use he makes of the relation that he regards these principles as valid.

The Principles of the Name-Relation

24-1. *The principle of univocality.* Every expression used as a name (in a certain context) is a name of exactly one entity; we call it the nominatum of the expression.

24-2. *The principle of subject matter.* A sentence is about (deals with, includes in its subject matter) the nominata of the names occurring in it.

24-3. *The principle of interchangeability* (or substitutivity).

This principle occurs in either of two forms:

a. If two expressions name the same entity, then a true sentence remains true when the one is replaced in it by the other; in our terminology (11-1b): the two expressions are interchangeable (everywhere).

b. If an identity sentence '. . . = - - -' (or '. . . is identical with - - -' or '. . . is the same as - - -') is true, then the two argument expressions '. . .' and '- - -' are interchangeable (everywhere).

The principle of univocality is, of course, applied only to a well-constructed language without ambiguities; its fulfilment may, indeed, be regarded as defining univocality in the sense of nonambiguity. (A language of this kind may, for instance, be an artificially constructed system or a modified English, where the ordinary ambiguities are eliminated, either by assigning to an ambiguous word only one of its usual meanings or by replacing it with several terms for the several meanings, e.g., 'probability$_1$', 'probability$_2$'.) The principle of subject matter is rather vague but sufficiently clear for our purposes. It is sometimes used for making the third principle plausible. And, indeed, if somebody accepts the first two principles, he will hardly reject the third. For, if $\mathfrak{A}_j$ and $\mathfrak{A}_k$ have the same nominatum and if the sentence $. . \mathfrak{A}_j . .$ says something true about this nominatum, then the sentence . . $\mathfrak{A}_k$ . . , saying the same about the same nominatum, must also be true. The form b of the third principle seems at first glance not to involve the name-relation at all. But it does so implicitly in the concept of identity sign or identity sentence. The following definitions of these concepts, it seems to me, are tacitly presupposed in 24-3b:

24-4. *Definitions.*

a. A predicator $\mathfrak{A}_l$ is an *identity expression* (for a certain type) $=_{\mathrm{Df}}$ for any closed expressions (names) $\mathfrak{A}_j$ and $\mathfrak{A}_k$ of the type in question, the full sentence of $\mathfrak{A}_l$ with $\mathfrak{A}_j$ and $\mathfrak{A}_k$ as argument expressions (i.e., $\mathfrak{A}_l(\mathfrak{A}_j, \mathfrak{A}_k)$ or $(\mathfrak{A}_j)\mathfrak{A}_l(\mathfrak{A}_k)$) is true if and only if $\mathfrak{A}_j$ and $\mathfrak{A}_k$ name the same entity.

b. $\mathfrak{S}_i$ is an *identity sentence* $=_{\mathrm{Df}}$ $\mathfrak{S}_i$ is a full sentence of an identity expression.

On the basis of these definitions, form b of the principle of interchangeability follows immediately from form a. Thus, granted the adequacy of these definitions, form b is just as plausible as form a. I think that Church[2] expresses the generally accepted conception when he says that the interchangeability of synonymous expressions, i.e., those which name the same entity, follows from "what seem to be the inevitable semantical and syntactical rules for '=' ".

We find an example of the method of the name-relation in Frege's procedure. His distinction between nominatum and sense will later be discussed in detail (§§ 28–30). He formulates the principle of interchangeability in the first form (24-3a) in this way:[3]

24-5. "The truth-value of a sentence remains unchanged if we replace an expression in it by one which names the same [entity]."

Another example of this method is Quine's analysis in [Notes]; he uses the terms 'designates' and 'designatum' in the sense of our 'names' and 'nominatum'. The principle of interchangeability in the second form (24-3b) is called by him the principle of substitutivity and is formulated in this way:

24-6. "Given a true statement of identity, one of its two terms may be substituted for the other in any true statement and the result will be true."[4]

This principle is not meant by Quine as a conventional rule for an identity sign in an artificial system but rather as an explicit formulation of a pro-

[2] [Review C.], p. 300. [3] [Sinn], p. 36. [4] [Notes], p. 113.

cedure which is customarily applied in the ordinary word language on the basis of the customary interpretation of the words. Quine distinguishes between the designatum of an expression and its meaning; this distinction is, as Church[5] has seen, in some respects very similar to Frege's.

The differences between the method of the name-relation and the method of extension and intension will later be discussed in detail. Here I wish to make only a few remarks in connection with the three principles. The concept of the extension of an expression is, as we shall see later, in some respects similar to the concept of its nominatum. Therefore, let us see to what extent analogues of the three principles hold for the concept of extension. The analogue of the principle of univocality holds; every designator has exactly one extension. The analogue of the principle of subject matter holds, too, but with restrictions. In general, a sentence containing a designator $\mathfrak{A}_j$ may be interpreted as speaking about the extension of $\mathfrak{A}_j$. However, it may be interpreted alternatively as speaking about the intension of $\mathfrak{A}_j$; and, as we shall see later, the latter interpretation is sometimes more appropriate. The decisive difference emerges with respect to the principle of interchangeability. For extensions, instead of the analogue of 24-3a, only the restricted principle 12-1 holds. It says that, if two expressions have the same extension, in other words, if they are equivalent, then they are interchangeable *in extensional contexts*. The principle 24-3b speaks about identity. However, on the basis of the method of extension and intension, we cannot simply speak of identity but must distinguish between identity of extension and identity of intension, in other words, between equivalence and L-equivalence. Therefore, instead of the one principle 24-3b for identity, we have in our method two principles, one for equivalence and the other for L-equivalence; these are 12-1 and 12-2.

§ 25. An Ambiguity in the Method of the Name-Relation

A predicator in a word language (e.g., 'gross' in German) or in a symbolic language (e.g., an abstraction expression in Quine's system) may be regarded as the name of a class but also as the name of a property. This shows an intrinsic ambiguity in the name-relation. Its consequences will be discussed later.

I shall now examine in more detail some features of the method of the name-relation, and especially try to show that the basic concept of this method is not so simple, clear, and unambiguous as it is usually supposed to be.

It seems generally to be assumed that, if we understand an expression,

[5] [Review Q.], p. 47.

we know at least to what kind of entities its nominatum belongs, and also in some cases which entity is the nominatum, although in other cases factual knowledge is required for this. For instance, if we understand German, then we know that the word 'Rom' is a name of the thing Rome, and that 'drei' is a name of the number Three. In the case of 'der Autor von Waverley' we know at least that it names, if anything, a (physical) thing; and if we have sufficient historical knowledge, we know that it is a name of the man Walter Scott. Analogously, in the case of 'die Anzahl der Planeten', we know at least that it names a number and, with the help of astronomical knowledge, we know that it names the number Nine. Generally speaking, given a full understanding of the language in question and, in particular, of some name in it and, in addition, all the factual knowledge relevant to the case in question, we should expect that there could be no doubt or controversy as to the nominatum of the name. However, it will now be shown that this is, in general, not the case.

Let G be a part of the German language, restricted to declarative sentences, with all dubious expressions and ambiguities eliminated (see explanation of 24-1) and, in particular, with the word 'gross' confined to its literal meaning concerning spatial extension. We imagine two logicians, L_1 and L_2, interested in the semantical analysis of G. Before they begin the theoretical analysis, they make certain in a practical way that they have the same interpretation or understanding of the language G; for instance, each agrees with any translation the other makes of a sentence of G into English. Then they begin their semantical analysis of G, according to the method of the name-relation based on the three principles (24-1, 2, 3). They examine the sentence in G: 'Rom ist gross'. They have no doubt and no disagreement as to its meaning; this is shown by the fact that both agree that its translation into English is: 'Rome is large'. Now they apply to the expressions in the given sentence the analysis in terms of the name-relation. Both agree that 'Rom' in G is a name of the thing Rome. But now suppose that with respect to the word 'gross' (or the phrase 'ist gross') the following controversy arises: L_1 says: "The sentence 'Rom ist gross' means that Rome belongs to the class Large. Hence it is about the thing Rome and the class Large. Therefore, according to the principle of subject matter, 'gross' is a name of the class Large; and hence, according to the principle of univocality, it cannot be a name of any other entity". Against this, L_2 says: "The given sentence means that Rome possesses the property Large. Hence it is about the thing Rome and the property Large. Therefore, according to the principle of subject matter, 'gross' is a name of the property Large; and hence, according to the principle of univocality, its nominatum cannot be any other entity; in particular, it cannot be the class Large."

We might perhaps try to reconcile the two logicians by pointing out that it does not really matter whether they say 'the sentence means that Rome belongs to the class Large' or 'the sentence means that Rome has the property Large', since these two assertions are both true and differ merely in their formulation. But, even if the two logicians were willing to agree with us on this point, the controversy concerning the nominatum of 'gross' would not be solved. Here, in distinction to the question concerning the whole sentence, they cannot simply agree that they are both right, that it does not matter whether they say that the nominatum is the class Large or that it is the property Large; for they agree in affirming the principles of the name-relation; therefore they must agree, according to the principle of univocality, that 'gross' (in G) can have only one nominatum. And, further, they agree that the class Large is not the same as the property Large; they agree generally in recognizing the distinction between a property and the corresponding class, as expressed, for instance, by 4-7 and 4-8.

Perhaps somebody will suggest to the two logicians that their insoluble controversy is due merely to the choice of an unsuitable object language; that a natural language like G, even after the elimination of obvious ambiguities, is not precise enough for univocal semantical analysis; and that, therefore, they should restrict their analysis to a well-constructed symbolic system with exact rules. I doubt whether the controversy is caused merely by the imperfections of G; but let us see what will result when the two logicians follow the suggestion. Let ML be the system constructed by Quine in [M.L.], and ML′ the system constructed out of ML by the addition, first, of the defined signs which Quine introduces in his book but does not count as parts of his system and, second, of a few nonlogical atomic matrices. The two logicians agree on the following interpretation of the system ML′: the primitive notation of ML is interpreted in accordance with Quine's explanations; on this basis the interpretations of the defined signs in ML′ are determined by their definitions; for the interpretation of the nonlogical atomic matrices, the following rule (similar to 1-2) is laid down:

25-1. *Rules of designation* (for ML′).

a. 'H*x*' is the translation of '*x* is a human thing'.

b. 'F*x*'—'*x* is a featherless thing'.

c. 'B*x*'—'*x* is a biped thing'.

'Thing' is here meant in the sense of 'physical thing'. ML′ is interpreted in such a way that things are taken as individuals in Quine's sense.[6] According to the rules 25-1, the three atomic matrices mentioned are fulfilled only by entities which are things, and hence both individuals and elements in Quine's sense.[7]

The two logicians agree not to take the signs 'H', 'F', and 'B', introduced by 25-1, as names, because it is obvious that otherwise they would immediately get into the same controversy concerning the nominata as they did with respect to the word 'gross' in G (compare the translations 4-2 and 4-3 of 'Hs'). They agree to take as names only those expressions which Quine calls closed terms, and among them especially the closed abstraction expressions, i.e., expressions of the form '$\hat{x}(\ldots x \ldots)$' without free variables.

Now the two logicians examine the following sentence in ML′: '$\hat{x}(Hx) \subset \hat{x}(Bx)$', which we call $\mathfrak{S}_1$. There is no doubt and no disagreement between them as to its meaning. They agree that, according to the rules of ML′,[8] $\mathfrak{S}_1$ is L-equivalent to '$(x)(Hx \supset Bx)$' and hence may be translated into 'for every x, if x is human then x is a biped' (see 4-4; we assume here that 'human' means as much as 'human thing', and 'biped' as 'biped thing'). However, as soon as they raise the question as to what is the nominatum of the abstraction expression '$\hat{x}(Hx)$', as it occurs in $\mathfrak{S}_1$, a controversy starts which is perfectly analogous to the earlier one with respect to 'gross' in G, in spite of the fact that we have here the exact system ML′. L_1 says: "We agree about the meaning of $\mathfrak{S}_1$, namely, that it is translatable as just stated; but it is likewise translatable into 'the class Human is a subclass of the class Biped' (4-6). Hence $\mathfrak{S}_1$ is about the class Human and the class Biped. Therefore, according to the principle of subject matter, '$\hat{x}(Hx)$' is a name of the class Human; hence, according to the principle of univocality, it cannot be a name of any other entity". L_2 replies: "Since $\mathfrak{S}_1$ is translatable as previously stated, it is likewise translatable into: 'the property Human implies (materially) the property Biped' (4-5). Hence $\mathfrak{S}_1$ is about the property Human and the property Biped. Therefore, according to the principle of subject matter, '$\hat{x}(Hx)$' is a name of the property Human; and hence, according to the principle of univocality, it cannot be a name of any other entity; in particular, it cannot be a name of the class Human." Since both logicians agree that the class Human is not the same as the property Human, they

[6] [M.L.], p. 135. [7] *Ibid.*, p. 131.

[8] In particular, the definitions D21 and D9 in [M.L.], pp. 185 and 133, apply here; note also the above remark on 25-1 concerning things.

must regard their statements concerning the nominatum of '$\hat{x}(Hx)$' as incompatible on the basis of the principle of univocality. In support of his statement, L_1 may point to the fact that Quine, the author of the system ML, says himself that the terms are names of classes,[9] that '$\subset$' is a sign of class inclusion,[10] and that the whole language deals with classes. L_2 may reply that he admits that the mode of speech used by Quine and by L_1 can be applied consistently; his point is that the same holds for the other mode of speech, which he uses. However, what makes the controversy insoluble is this: The divergence between L_1 and L_2, which is at the start nothing but a difference in the mode of speech, namely, between the translations of $\mathfrak{S}_1$ in terms of classes and in terms of properties, leads, on the basis of the principles of the name-relation, to two statements which are incompatible, namely, those concerning the nominatum of '$\hat{x}(Hx)$'.

Now L_1 discovers a new way which, he thinks, must lead to an unambiguous solution of the puzzling problem. Since the difference between classes and properties has its root in the difference of the identity conditions, an identity sentence $\mathfrak{A}_i = \mathfrak{A}_j$ in ML′ should be analyzed where $\mathfrak{A}_i$ and $\mathfrak{A}_j$ are abstraction expressions; by determining the truth-condition of this sentence, we should be able to see, he thinks, whether the two expressions $\mathfrak{A}_i$ and $\mathfrak{A}_j$ are names of classes or of properties. Therefore, he proposes to examine the following sentence in ML′: '$\hat{x}(Hx) = \hat{x}(Fx \bullet Bx)$', which we call $\mathfrak{S}_2$. There is again complete agreement between the two logicians as to the meaning of this sentence. They agree that, according to the rules of ML′,[11] the sentence $\mathfrak{S}_2$ is L-equivalent to '$(x)(Hx \equiv Fx \bullet Bx)$' and hence, on the basis of the biological fact 3-6, $\mathfrak{S}_2$ is true. Further, both agree that the two classes in question are, in fact, identical (see 4-7), while the two properties are not (see 4-8). Now L_1 argues as follows: "The identity sentence $\mathfrak{S}_2$ can only refer to the two classes; for, if it referred to the two properties, it would be false because they are nonidentical". L_2 replies: "You, like the author of the system, take '$=$' as a sign of identity of classes. I admit that this is in accordance with the rules of the system ML′. But then, '$=$' may just as well be called a sign of equivalence of properties (like '$\equiv$' in S_1; see remark on 5-3). And since the two properties in question, though not identical, are indeed equivalent (see 5-5), $\mathfrak{S}_2$ is also true on the basis of this analysis, which interprets the two abstraction expressions as names of properties."

L_1 will perhaps ask whether the character of '$=$' in the system ML′ as

[9] [M.L.], p. 119.

[10] *Ibid.*, p. 185.

[11] See, in particular, the definitions D10 and D9 in [M.L.], pp. 136 and 133.

a genuine sign of identity and not merely a sign of equivalence, like '≡' in S_1, is not assured by the fact that ML′ contains a principle of interchangeability (called principle of substitutivity of identity[12]). To this, L_2 will give a negative answer. Interchangeability on the basis of '≡' holds likewise in S_1 (see 12-3a); thus, in this respect also, '=' in ML′ is like '≡' (between predicators) in S_1. It is true that general interchangeability on the basis of '≡' does not hold in some systems, for example, in S_2; but it holds in all extensional systems (12-3a). Thus the effect of the principle of interchangeability in ML′ (and ML) is simply to make ML′ (and ML) an extensional language like S_1; the principle prevents the introduction into ML′ of intensional predicators or connectives, for instance, of a sign of logical necessity (like 'N' in S_2, see § 11, Example II). But it does not prevent in any way the interpretation of abstraction expressions in ML′ (or ML) as names of properties.

Now let us draw the conclusion from our examination of the controversy between the two logicians. Note that this controversy is not an instance of the well-known multiplicity of interpretations, that is, of the fact that for a given logical system (calculus) there are, in general, several interpretations, all of them in accordance with the rules of the system. L_1 and L_2 apply the same interpretation to their object language G, and then likewise to the language system ML′. Even when L_1 says that the sign '=' in ML′ is a sign of identity of classes while L_2 says that it is a sign of equivalence of properties, this does not show a difference in interpretation but merely a difference in the choice of semantical terms used for describing one and the same interpretation; for equivalence of properties is just the same as identity of classes (or, speaking more exactly, 'the properties expressed by two predicators are equivalent' and 'the corresponding classes are identical' are L-equivalent sentences in M). That L_1 and L_2 apply the same interpretation to ML′ (as well as to G) means that to any given sentence in ML′ they attribute the same meaning or, in other words, the same truth-condition. The decisive point is rather this: In spite of their agreement in the interpretation, it is possible for L_1 and L_2 to maintain different conceptions as to what are the nominata of the names occurring—conceptions which are incompatible with each other, though each is consistent in itself. This shows, it seems to me, that the method of the name-relation involves an intrinsic ambiguity, inasmuch as the fundamental term of this method, namely, 'is a name of',

[12] Quine, [M.L.], § 29, *201; for the corresponding principle with respect to the word language, see above, 24-6.

is ambiguous, although it is generally believed to be quite clear and unambiguous. This is not to say that, in general, a logician uses these terms ambiguously, but only that several logicians may use them in different ways. For instance, L_1 uses the method consistently and unambiguously, and so does L_2. The trouble is that, if one logician thinks that the results which he has found on the basis of his conception must be accepted by everybody else, he is mistaken, because it may be that the results do not hold for another conception of the name-relation.

We have discussed the ambiguity only with respect to predicators, where either classes or properties may be taken as nominata. Analogously, for a designator of another kind, either its extension or its intension may be taken as its nominatum. Thus there are, in fact, many more than two ways for using the method of the name-relation. And the multiplicity of ways is, further, considerably increased by the fact that some logicians take some predicators as names of classes and other predicators of the same type as names of properties (see § 26); and that some logicians even take the same expression as a name of an extension in one context and in another as a name of an intension (for example, Frege, see below, §§ 28, 29). For the present, it will suffice to point out the great multiplicity of different ways of using a method of the name-relation, in other words, the many different senses in which the term 'name' or similar terms are used. Some of these ways will be discussed later in order to show the complications which they involve.

§ 26. The Unnecessary Duplication of Names

Many systems have different names for properties and for the corresponding classes. This is discussed with respect to examples from the system of *Principia Mathematica*. Analyzing these names by the method of extension and intension, we find that a name for the property Human and a different name for the class Human have not only the same extension but also the same intension. Therefore, the duplication of names to which the method of the name-relation leads is superfluous.

Another consequence of the customary way of using the method of the name-relation will now be discussed. The principle of subject matter (24-2) says that if a sentence contains a name of an entity, then it says something about this entity. And the method is usually conceived in such a way that, conversely, if a sentence is intended to be about a certain entity, then it must contain a name of this entity. Then it follows, in virtue of the principle of univocality (24-1), that, in order to speak about two different entities, we have to use two different expressions as their names.

On the basis of the method of extension and intension, on the other hand, the situation is quite different. A designator is here regarded as having a close semantical relation not to one but to two entities, namely, its extension and its intension, in such a way that a sentence containing the designator may be construed as being about both the one and the other entity. Thus here, if a sentence is intended to speak about an entity which is an extension, an expression is needed whose extension is that entity; and if we wish to speak about an entity which is an intension, an expression is needed whose intension is that entity. Therefore, in order to speak first about a certain intension and then about the corresponding extension, this method requires only one expression, while the method of the name-relation would require two and hence lead to an unnecessary duplication in symbolism.

This duplication can best be made clear in the case of predicators. The method of extension and intension needs only one predicator to speak both about a certain property and about the corresponding class. The method of the name-relation in its customary form, however, needs for this purpose two different expressions, a property name and a class name. As an example, let us take the symbolic system PM constructed by Whitehead and Russell in [P.M.]; PM includes not only the primitive signs but also the (logical) signs introduced by the definitions as given by the authors. Let PM′ consist of PM and, in addition, a few nonlogical predicators or atomic matrices. Let PM′ be interpreted in the following way: The primitive logical signs are interpreted in accordance with the explanations of the authors of [P.M.]; the interpretations of the defined signs are then determined by their definitions; the nonlogical signs are interpreted by 25-1 as a rule of designation for PM′.

The system PM′ uses different expressions as names for properties (construed as propositional functions) and as names for classes. Take, as examples, the following four statements concerning two pairs of expressions in PM′:

26-1. '$H\hat{x}$' is a name of the property Human.
26-2. '$\hat{x}(Hx)$' is a name of the class Human.
26-3. '$F\hat{x} \bullet B\hat{x}$' is a name of the property Featherless Biped.
26-4. '$\hat{x}(Fx \bullet Bx)$' is a name of the class Featherless Biped.

[For the present discussion we may leave aside the fact that Russell does not assume that there are classes as separate entities, in addition to properties; he introduces class expressions by contextual definitions on the basis of property expressions. The problem of this and the converse reduction will be discussed later (§ 33). For our present problem it is sufficient that an author speaks in his metalanguage both of properties (qualities, propositional functions of degree one) and of classes (distinguished in the customary way); that he uses in his object language two different kinds of expressions; and that he declares that those of the first kind are meant as expressions of properties and those of the second kind as expressions of classes.]

The four statements given express results of a semantical analysis of certain expressions in PM′, according to the method of the name-relation. If, instead, we analyze PM′ by the method of extension and intension, we arrive at the following results, which contain counterparts of the earlier results, supplemented by new ones. Instead of 26-1, we have here:

26-5. The intension of '$H\hat{x}$' is the property Human.

To this statement, however, another statement is added, which follows from it:

26-6. The extension of '$H\hat{x}$' is the class Human.

Instead of 26-2, we have here:

26-7. The extension of '$\hat{x}(Hx)$' is the class Human.

To this we add:

26-8. The intension of '$\hat{x}(Hx)$' is the property Human.

While 26-6 follows directly from 26-5, the same is not true for 26-8 and 26-7; every intension uniquely determines an extension, but the converse does not hold. Statement 26-8 is based, rather, on the rule 25-1a and the circumstance that, according to the rules of PM′, the sentence '$(y)[y \epsilon \hat{x}(Hx) \equiv Hy]$' is L-true in PM′. The results corresponding to 26-3 and 26-4 are, of course, analogous.

Thus the outcome, from the point of view of our method, is that the two expressions '$H\hat{x}$' and '$\hat{x}(Hx)$' in PM′ have the same extension and also the same intension. Therefore, it is unnecessary to have both forms in the system. The two expressions are, in a certain sense, L-equivalent predicators. It is true that one of them cannot simply be replaced by the other; this is the effect of certain restricting rules concerning the two kinds of predicators. First, there is the following unessential difference, which is merely an accidental syntactical feature of the systems PM and PM′. The rules require that an argument expression for a predicator of the first kind (e.g., 'H' or '$H\hat{x}$') succeeds it (resulting in 'Hs'), while one for a predicator of the second kind precedes it with a copula 'ϵ' interposed (e.g., '$s \epsilon \hat{x}(Hx)$'). Another difference is more important. It concerns identity sentences built with '='. Consider the following two sentences as examples:

26-9. '$\hat{x}(\mathrm{H}x) = \hat{x}(\mathrm{F}x \bullet \mathrm{B}x)$'.
26-10. '$\mathrm{H}\hat{x} = \mathrm{F}\hat{x} \bullet \mathrm{B}\hat{x}$'.

According to the explanation given in [P.M.], the sentence 26-9 says that the two *classes* in question are identical; hence this sentence is true (see 4-7). On the other hand, the sentence 26-10 says that the two *properties* in question are identical; hence this sentence is false (see 4-8). Thus, 26-9 is in notation and meaning just like a sentence in ML′ previously discussed ($\mathfrak{S}_2$ in § 25). Likewise, its L-equivalence to '$(x)(\mathrm{H}x \equiv \mathrm{F}x \bullet \mathrm{B}x)$' holds for PM′. Therefore, the contention of L_2 that '=' in 26-9 is like '≡' in S_1 (or S_2) and, hence, is simply a sign of equivalence applies here as well. On the other hand, '=' in 26-10 is a sign of identity or L-equivalence of properties; it is therefore, in distinction to '=' in 26-9, a nonextensional sign. (This is recognized by Whitehead and Russell.)[13] Hence it cannot correspond to any sign in the extensional language S_1; but it corresponds exactly to the modal sign '≡' in S_2, which will be introduced later (see 39-6; accordingly, the false sentence 26-10 is L-equivalent to 42-2bA without the sign of negation). Thus the method of extension and intension by no means overlooks the difference between 26-9 and 26-10. On the basis of this method, in distinction to the method of the name-relation, the first components in the two sentences (i.e., the predicators '$\mathrm{H}\hat{x}$' and '$\hat{x}(\mathrm{H}x)$') are equalized in certain respects, and so are the second components. Nevertheless, the difference is preserved because the occurrences of '=' in 26-9 and in 26-10 are here construed as having different meanings. The first is interpreted as a sign of equivalence or, in other words, of identity of extensions; the second as a sign of L-equivalence or, in other words, of identity of intensions.

We see that the situation with respect to the two methods under discussion is this: At the beginning, there is merely a difference of procedure in describing the semantical features of given language systems. The customary method does it in terms of nominata; our method does it, instead, in terms of extensions and intensions. At first glance, one might think that both methods were neutral with respect to the structure of the language systems, in the sense that either method is as applicable to any system as the other. If so, the choice of the one or the other method of semantical analysis would not have any effect upon the choice of a structure for a system to be constructed. However, this is not so. According to

[13] [P.M.], I, 84.

the first method, the two expressions '$H\hat{x}$' and '$\hat{x}(Hx)$' are said to have different nominata; and this circumstance is then naturally regarded as justification for the decision to incorporate both expressions into the system, as is done in the system PM′. According to the second method, on the other hand, the two expressions are said to have the same extension and the same intension. This leads to the view that the inclusion of both would be an unnecessary duplication, and hence to the decision to construct the system in such a way that it contains, instead of those two expressions, only one, as in the systems S_1 and S_2 (and in many systems constructed by other logicians[14]). Corresponding to the two expressions in PM′, S_1 and S_2 have the one predicator '$(\lambda x)(Hx)$' (of course, either of the two notations in PM′ could be taken, instead, just as well). That we could do in previous examples (e.g., 3-8) without lambda-expressions was merely due to the simplicity of the examples. In general, an identity sentence for classes in PM′ (like 26-9) will be translated into S_1 and S_2 in the form '$(\lambda x)(\ldots) \equiv (\lambda x)(---)$', and the corresponding identity sentence for properties (like 26-10) will be translated into S_2 in the form '$(\lambda x)(\ldots) \equiv (\lambda x)(---)$', with the same two lambda-expressions as the first sentence.

Our conclusion that the duplication of predicators in PM and PM′ is unnecessary holds likewise for systems which use two different kinds of operators for class abstraction (e.g., '$\hat{x}(\ldots x \ldots)$') and for functional abstraction,[15] that is, formation of abstraction expressions for properties, here construed as propositional functions (e.g., '$(\lambda x)(\ldots x \ldots)$'). Here again, if the same matrix '$\ldots x \ldots$' occurs as scope in both expressions, they have the same extension and the same intension; however, they have different conditions of identity. Thus they are analogous to '$\hat{x}(Hx)$' and '$H\hat{x}$', respectively, in PM′.

Since the choice of a semantical method and the choice of a form of language are interconnected, we may also reason in the inverse direction: our preference for a language structure may influence our preference for one of the two semantical methods. If a language system with only one kind of predicator is, in fact, not only as effective (for the purposes of both mathematics and empirical science) as a system with two kinds like PM′

[14] That it is unnecessary to have special class expressions in addition either to simple predicator signs and their combinations or to property expressions has already been seen by several logicians. Concerning the historical development of this insight and concerning the possibility of a form of language without special class expressions, see [Syntax], §§ 38 and 37. The discussion in the present book confirms this conception by basing it on a more general conception, namely, that of the method of extension and intension for designators in general.

[15] See, for instance, Church, [Dictionary], p. 3.

but also simpler and hence more convenient, then I think the method of the name-relation must be regarded as at least misleading, if not inadequate.

§ 27. Names of Classes

A name for a class must be introduced by a rule which refers to exactly one property; otherwise, the meaning of the new sign and of the sentences in which it occurs is not uniquely determined. This shows that a semantical rule for a sign determines primarily its intension; only secondarily, with the help of relevant facts, its extension. The customary use of different kinds of variables for properties and for classes is shown to be as unnecessary as that of different names. The duplication of names and variables on the first level leads to a still greater multiplication of names and variables on higher levels. The concepts of mathematics can be defined without the use of special class expressions and class variables. This is shown by definitions of '2' and of 'cardinal number'.

We have seen in the preceding section that those expressions in the system PM′ which are regarded as names of certain classes by the authors of the system do not only have these classes as their extensions but, at the same time, have certain properties as intensions (see 26-8). Here the question might be raised as to whether it could not happen in some system that a predicator has only an extension, not an intension; in other words, that it refers to a class without referring to any of those properties which have that class as an extension. I think that this is not possible in a semantical system, that is, in a system whose interpretation is completely given. To begin with, it is not possible to refer to a class without referring to at least one of the corresponding properties. This holds, even if the class is specified by an enumeration of its members, e.g., by a phrase like 'the class of the individuals a, b, and c', or in the symbolic language S_1: '$(\lambda x)[(x \equiv a) \vee (x \equiv b) \vee (x \equiv c)]$'. This predicator does not lack an intension; it is the property of being (identical with) either a or b or c. The feeling which we might have, that this is not a property in the same sense as properties like Blue or Human, is right; it is (if 'a', 'b', and 'c' are interpreted as L-determinate constants for positions in an ordered domain, § 19) a positional, not a qualitative, property; in our earlier terminology (§ 22), it is an L-determinate property; but, in any case, it is an intension.

One might perhaps think a class name without an intension could be introduced into a system by stipulating that it is to be a name for the class which such and such equivalent properties have in common; this reference to several properties would have the effect that none of them would be *the* intension of the name. Consider, for instance, the following as a semantical rule for the class name 'K' in S_1:

27-1. 'K' is to be a name at once for the class Human and for the class Featherless Biped, which is the same class.

This rule does not involve an inconsistency, since the classes mentioned are indeed identical (see 4-7). However, it is not sufficient as a semantical rule for 'K'; the interpretation of 'K' or, in ordinary words, its meaning, is not completely given by 27-1 but merely confined to certain possibilities. It is true that this rule, together with rules for the other signs in S_1 and knowledge of the relevant facts, is sufficient to determine the truth-value of any sentence in S_1 in which 'K' occurs. For instance, 'Ks' is found to be true in S_1 on the basis of the historical facts which make the two sentences 'Hs' and 'Fs • Bs' true. The decisive point is that, although the truth-values, the extensions, of the sentences containing 'K' are determined, their intensions are, in general, not. For instance, it remains undetermined what proposition is expressed by 'Ks'; is it the same as that expressed by 'Hs', or by 'Fs • Bs', or by their disjunction, or their conjunction? These are four different propositions. To express it in other terms, the given K-rule (27-1), together with the rules for other signs, does not suffice for the application of the L-concepts to the sentences containing 'K'. For instance, it is not determined whether 'Ks $\equiv$ Hs' is L-true or F-true. Therefore, strictly speaking, on the basis of the K-rule and the other rules we cannot *understand* sentences like 'Ks' or 'Ks $\equiv$ Hs', although we can establish their truth-values. The reason for the objection here raised against the K-rule is not the fact that it introduces 'K' as a name for a class, but rather the fact that it does not do this by reference to exactly one property. In contradistinction to 27-1, the following would be a complete semantical rule for 'K':

27-2. 'K' is to be a name for the class Human.

For this would say the same as: ''K' is to be a name for the class which is the extension of the property Human'; and this, in turn, may be understood as saying: ''K' is to be a sign whose intension is the property Human; therefore, its extension is the class Human.' The first part of this last sentence would suffice as a rule; the second part ('therefore . . .') is a semantical statement following from the rule. This shows that *the semantical rule for a sign has to state primarily its intension; the extension is secondary,* in the sense that it can be found if the intension and the relevant facts are given. On the other hand, if merely the extension were given, together with all relevant facts, the intension would not be uniquely determined.

We have seen in the preceding section how the method of the name-relation leads to the use of two kinds of predicators within the same type (for example, level one and degree one). On the basis of this method, especially of the principle of subject matter, this duplication of predicators is regarded as necessary if we wish to speak both about classes and about properties. An analogous situation arises with respect to *variables*. For speaking about particular entities, names are used; and thus the method leads to class names and property names. On the other hand, for speaking about entities of some kind in a general way, variables are used; thus here the method of the name-relation leads to the introduction of two kinds of predicator variables for the same type; the values of variables of the first kind are classes, the values of those of the second are properties. Thus, for example, the system PM uses 'α', 'β', etc., as class variables and 'ϕ', 'ψ', etc., as variables for properties (propositional functions). From the point of view of the method of extension and intension, this duplication is analogous to that of closed predicators and just as superfluous. In the system PM′, '$\hat{x}(\mathrm{H}x)$' is a value expression for 'α'. We have seen that, on the basis of our method, '$\hat{x}(\mathrm{H}x)$' has not only an extension, namely, the class Human (see 26-7), but also an intension, the property Human (see 26-8). Therefore, not only does the class Human belong to the value extensions of 'α' according to 10-1, but it is also the case that the property Human belongs to the value intensions of 'α' according to 10-2. But exactly the same holds for 'ϕ' because of 26-6 and 26-5, since '$\mathrm{H}\hat{x}$' is a value expression for 'ϕ'. Thus both kinds of variables have the same value extensions, namely, classes of individuals, and the same value intensions, namely, properties of individuals. Therefore, the duplication of variables is as unnecessary as that of closed predicators. It is sufficient to use one kind of variable for the predicator type in question; their value extensions are classes, their value intensions are properties (see § 10). Therefore, they serve for speaking in a general way both about classes and about properties. [Thus, for instance, with respect to the examples in § 10 preceding 10-1, sentences of both the forms (ii) and (iii) are translated into a symbolic language with the help of the same variable 'f' in the form '$(\exists f)(\ldots f \ldots)$'.]

The situation with respect to variables of other kinds is theoretically the same but practically different; while many logicians use different variables for classes and for properties, it seems that hardly anybody proposes to use different variables for propositions and for truth-values, or different variables for individuals and for individual concepts. Thus our method does not deviate here from the customary procedure.

If the reasoning on the basis of the method of the name-relation, which leads to the use of two kinds of predicators within the simplest type, is carried to higher levels, then it results in an immense multiplication of predicators of the same type. From our point of view this multiplication is as unnecessary as the duplication with which it starts. For the sake of simplicity, let us restrict the discussion to predicators of degree one, that is to say, let us speak only of classes and properties, leaving relations aside. If on the first level a distinction is made between names of classes and names of properties, then, on the second level, four kinds of predicators must be distinguished, namely:

names of classes of classes
names of properties of classes
names of classes of properties
names of properties of properties

To form examples in the system PM, let us start with the following matrix, which contains the class variable 'α' as the only free variable:

$$`(\exists x)(\exists y)[\sim (x = y) \bullet (z)(z \,\epsilon\, \alpha . \equiv : z = x . \vee . z = y)]'.$$

As shorthand for this in the subsequent examples, let us simply write '$. . \alpha . .$'. This matrix says that the class α has exactly two members, or, as we may say for short, that α is a pair-class. Let '$. . \phi . .$' be taken as shorthand for that matrix in PM which is analogous to the one mentioned but which contains the property variable 'ϕ' instead of 'α' (that is to say, '$z \,\epsilon\, \alpha$' is replaced by 'ϕz'). Hence, '$. . \phi . .$' says that there are exactly two individuals which have the property ϕ, or, as we may say, that ϕ is a pair-property. Now let us examine the following four expressions in PM:

(i) '$\hat{\alpha}(. . \alpha . .)$',
(ii) '$. . \hat{\alpha} . .$',
(iii) '$\hat{\phi}(. . \phi . .)$',
(iv) '$. . \hat{\phi} . .$',

where the dots indicate the matrices just described. Expression (i) is a name of the class Pair-Class and hence belongs to the first of the four kinds of predicators on the second level mentioned above; (ii) is a name of the property Pair-Class and hence belongs to the second kind; (iii) is a name of the class Pair-Property and hence belongs to the third kind; (iv) is a name of the property Pair-Property and hence belongs to the fourth kind. The nominatum of (i), that is, the class of all classes which have exactly two members, is in PM taken as the cardinal number Two, and therefore '2' is introduced as abbreviation for (i). The expressions (ii), (iii), and (iv) do not, it seems, actually occur in the book [P.M.], but they are formed according to the rules of the system PM. The four expressions belong to the same type; they are predicators of level two and degree one. If we were to construct, on the basis of our method of extension and intension, a system with a predicator variable 'f', then it would contain, instead of the four expressions of PM, only one, namely, '$(\lambda f)(\ldots f \ldots)$'.

The multiplication of kinds of predicators on the basis of the method of the name-relation increases with higher levels. On the level n, there are 2^n different kinds of predicators within the same type. They are supposed to be required as names of 2^n kinds of entities. On the basis of our method, there is only one kind of predicator in each type; and the 2^n corresponding predicators in the other method are here replaced by one.

On the basis of our method, all the mathematical concepts can be defined in a way that is analogous to that in [P.M.] except that no special class expressions and class variables are used. Let us suppose that S is a system which contains not only individual variables but also variables for which predicators of various levels can be substituted, say 'f' and 'g' as variables of level one and 'm' and 'n' as variables of level two. Then, for example, the cardinal number Two can be defined in S as a property of properties as follows:

27-3. '2' for '$(\lambda f)[(\exists x)(\exists y)[\sim(x \equiv y) \bullet (z)(fz \equiv (z \equiv x) \vee (z \equiv y))]]$'.

It is true that a certain requirement of extensionality must be fulfilled by any explicatum for the concept of cardinal number in order to be adequate. However, it is not necessary to require that the cardinal numbers be extensions; it is sufficient to require that any statement attributing a cardinal number to a given property (or class) be extensional. This requirement is also fulfilled by our method, because the cardinal numbers are here defined as properties of properties which are extensional. That, for example, 2 as defined by 27-3 is an extensional property of properties is not explicitly stated in the definition, but it is seen from the fact that the following sentence is provable with the help of the definition 27-3:

$$'(f)(g)[(f \equiv g) \supset (2(f) \equiv 2(g))]'.$$

The general concept of cardinal number can likewise be defined in the system S without the use of special class expressions. While Russell explicates cardinal numbers as classes of classes, Frege takes them as classes of properties. Since we wish to take them as properties of properties, we may follow Frege's procedure half the way. We say, like Frege,[16] that the property f is equinumerous to the property g (in symbols: 'Equ(f, g)') if there is a one-to-one correlation between those individuals which have the property f and those which have the property g. Then we define the cardinal number of the property f as the property (of second level) Equinumerous To f:

27-4. 'Nc'f' for '$(\lambda g)[\text{Equ}(g, f)]$'.

[Frege takes as definiens not 'the property Equinumerous To f', but 'the extension of the property Equinumerous To f', which means the same as 'the class Equinumerous To f'. Now it is interesting to see that Frege adds to this definition a footnote (*op. cit.*, p. 80) which says: "I believe that instead of 'extension of the property' we might say simply 'property'. But two objections would be raised: . . . . I am of the opinion that both these objections could be removed; but that might lead here too far." Thus Frege considers here the simpler procedure which we now adopt. He seems to regard it as feasible but does not pursue it any further. In his later work[17] he again defines cardinal number in the way stated above, without even mentioning an alternative possibility. His chief reason for regarding cardinal numbers as classes of properties rather than as properties of properties seems to be his view[18] that cardinal numbers are independent entities, in combination with his general conception that classes are independent entities, while properties are not. However, I find his reasoning on this question not quite clear and far from convincing.]

Finally, we define, like Frege,[19] 'n is a cardinal number' (in symbols: 'NC(n)') by 'there is a property f such that n is the cardinal number of f':

27-5. 'NC' for '$(\lambda n)[(\exists f)(n \equiv \text{Nc}`f)]$'.

Suppose that the properties f and g are equinumerous. Frege shows on the basis of his definitions that in this case the cardinal number of f is equal to that of g. The latter statement is interpreted by him as saying that the class Equinumerous To f is the same as the class Equinumerous To g. Thus he explicates equality of numbers as identity. Here our defini-

16 [Grundlagen], pp. 73–79, 83–85.

17 [Grundgesetze], I, 57.

18 [Grundlagen], pp. 67–72.

19 *Ibid.*, p. 85.

tion 27-4 may seem to involve a difficulty, because, even if *f* and *g* are equinumerous, the property Equinumerous To *f* need not be the same as the property Equinumerous To *g*. However, although these two properties, which in our method are regarded as cardinal numbers, are not identical, they are equivalent (in the sense of 5-3; see the example 5-5). Thus the difficulty disappears if we explicate equality of numbers as equivalence rather than as identity and hence symbolize it by '≡'. Thus, for example, the sentence

'the number of planets = 9'

would be translated into the system *S* as follows, if we take 'P' as predicator for the property Planet:

27-6. 'Nc'P ≡ 9'.

(The definition of '9' is, of course, analogous to that of '2' in 27-3.)

We have said that we explicate cardinal numbers as properties of second level, in contrast to Frege and Russell, who take them as classes of second level. But this formulation is a concession to the customary view based on the name-relation, according to which a predicator is a name either of a class or of a property and cannot refer to both of them at once. According to the method of extension and intension, it would be more adequate to say that we introduce cardinal number expressions as predicators of second level and that these predicators have as intensions properties of second level and as extensions classes of second level. Thus, for example, '2' is a predicator of second level; its intension is the property (of second level) Two, which we might call the number intension Two or the number concept Two; and its extension is the class (of second level) Two, which we might call the number extension Two. Since the sentence 'Nc'P ≡ 9' is true but not L-true, the predicators 'Nc'P' and '9' are equivalent but not L-equivalent. Therefore, the number extension The Number Of Planets is the same as the number extension Nine, while the number intension The Number Of Planets is not the same as, but equivalent to, the number intension Nine. Thus we see that in our method, too, as in those of Frege and Russell, equality of numbers can be regarded as identity of certain entities, not of number intensions but of number extensions.

In this way the whole system of mathematics constructed on the basis of logic by Frege and Russell can be reconstructed in a simpler form without the use of class expressions distinct from property expressions and of class variables distinct from property variables.

§ 28. Frege's Distinction between Nominatum and Sense

Frege distinguishes for any name between its nominatum, i.e., the object named, and its sense, i.e., the way in which the object is given by it. We see from Frege's discussion that his concept of nominatum fulfils the principles of the name-relation stated earlier (§ 24); thus his method of semantical analysis is a particular form of what we call the method of the name-relation. According to Frege, the nominatum of an isolated sentence is its truth-value, and its sense is the proposition expressed by it. However, if the sentence stands in an oblique (i.e., nonextensional) context, then its nominatum is that same proposition.

Frege[20] has made a very interesting distinction between the nominatum of an expression and its sense.[21] This distinction will now be explained and then, in the next section, compared with our distinction between extension and intension. It will be seen that in some respects there is a close similarity between the two kinds of distinctions; and it was, indeed, Frege's pair of concepts that first suggested to me the concepts of extension and intension as applied to designators in general. On the other hand, we shall find differences between the two conceptions, based chiefly upon the fact that Frege's conception is a particular form of what I have previously called the method of the name-relation.

The purpose of Frege's paper, described here in modern terminology, is to carry out a semantical analysis of certain kinds of expressions in the ordinary word language and to propose, examine, and apply semantical concepts as instruments for this analysis. His discussions seem to me of great importance for the method of logical analysis; but, like his other works, this paper has not found the attention it deserves. Except for Russell, [Denoting], who has discussed Frege's analysis in detail but rejected most of it, Frege's paper seems to have been neglected for about half a century, until Alonzo Church[22] began, several years ago, to point

[20] [Sinn].

[21] I list here the English terms which I shall use as *translations of Frege's terms*, following, in most cases, Russell, [Denoting], and Church (see n. 22). 'Ausdrücken' is translated into 'to express' ('to connote' might perhaps also be taken into consideration, in analogy to 'to denote', although it often has in ordinary usage a quite different sense which concerns not the designative meaning component but other ones, especially the associative and emotive); 'Sinn'—'sense' (so Church; Russell uses 'meaning'; 'connotatum' or 'connotation' might also be considered); 'bezeichnen'—'to be a name of' or 'to name' (Russell and Church: 'to denote'; see the remark on the ambiguity of this term in n. 1, § 24); 'Bedeutung'—'nominatum' (Russell and Church: 'denotation'); 'Begriff'—'property' (Frege uses 'Begriff' for attributes of degree one only; for attributes in general he uses the phrase 'Begriff oder Beziehung'); 'Gedanke'—'proposition' (see Church's justification for this translation, [Review Q.], p. 47); 'gewöhnlich (Rede, Bedeutung, Sinn)'—'ordinary'; 'ungerade (Rede, Bedeutung, Sinn)'—'oblique'; 'Gegenstand'—'object'; 'Wertverlauf'—'value distribution'; 'Behauptungssatz'—'(declarative) sentence'.

[22] In reviews in the *Journal of Symbolic Logic*, V (1940), 162, 163; VII (1942), 101; see also an abstract of a paper of his, *ibid.*, VII, 47; further, more in detail, in [Dictionary] article, "Descriptions", [Review C.], and [Review Q.].

out repeatedly the importance of Frege's conception, defending its basic idea while beginning to develop further the details of its application.

Frege's distinction between nominatum and sense is made in the following way: Certain expressions are names of objects (this term is to be understood in a wide sense, including abstract, as well as concrete, objects) and are said to name ('bezeichnen') the objects. From the *nominatum* of an expression, that is, the object named by it, we must distinguish its *sense;* this is the way in which the nominatum is given by the expression. This is illustrated by the following example:

28-1. The two expressions 'the morning star' and 'the evening star' have the same nominatum.

This holds because both are names of the same thing, a certain planet; in other words, the following is a true statement of an astronomical fact:

28-2. The morning star is the same as the evening star.

On the other hand, the following holds:

28-3. The expressions 'the morning star' and 'the evening star' do not have the same sense.

The reason for this is that the two expressions refer to their common nominatum, that planet, in different ways. If we understand the language, then we can grasp the sense of the expressions; for instance, we are then aware that the sense of 'the morning star' is the same as that of the phrase 'the body which sometimes appears in the morning before sunrise in the eastern sky as a brightly shining point'. The nominatum is not, however, given by the sense but only, as Frege puts it, illuminated from one side ("einseitig beleuchtet"). To find the result 28-1, more is required than merely to understand the sense of the expressions (namely, observation of facts).

After having explained the distinction in a general way, Frege proceeds to apply it to sentences. In a (declarative) sentence we express a proposition ('Gedanke'). Is the proposition expressed by a sentence its sense or its nominatum? By a long and careful analysis, Frege arrives at the following two results:

28-4. The (ordinary) sense of a sentence is the proposition expressed by it.
28-5. The (ordinary) nominatum of a sentence is its truth value.

These are the results for ordinary cases; they hold, in particular, for any isolated sentence, that is, one which is not a part of a larger sentence; the exceptions will be discussed later. For our purposes the most important question to be raised here concerns the method by which Frege arrives at these two results (and at the exceptions to them). They are clearly not meant simply as conventions, as, so to speak, part of the definitions of the terms 'sense' and 'nominatum'. If this had been Frege's intention, he probably would have chosen a simple general rule not complicated by exceptions. It becomes clear from his discussion that the situation is otherwise. Frege assumes that he knows quite clearly what he means by 'sense' and 'nominatum', that is, that he knows the way in which he intends to use these terms. On the basis of this knowledge, he investigates how these terms apply to various kinds of expressions. Thereby he discovers objective results, and these he reports as he finds them, whether they are simple or complicated. For the reader, however, it is not so clear as for Frege himself what is to be understood by his two terms. The preliminary explanations which he gives are certainly not sufficient to lead to the results, or even to make them plausible. The nominatum of an expression, for instance, is explained as that of which the expression is a name. This explanation, however, by no means succeeds in making the result 28-5 plausible. I think any unprepared reader would be inclined to regard a sentence as a name of a proposition rather than as a name of a truth-value—if, indeed, he is at all willing to regard a sentence as a name of anything. Another explanation for 'nominatum' which Frege gives is that a sentence is about the nominata of the expressions occurring in it (we have previously called this the principle of subject matter, 24-2). But this explanation, it seems to me, does not make 28-5 any more plausible. Take as an example the false sentence 'Hw' (see rules 1-1 and 1-2) as part of '$\sim$Hw'. (According to Frege, this is an ordinary case, that is to say, 28-4 and 28-5 also hold for 'Hw' in this context.) The question here is whether the nominatum of 'Hw' as part of '$\sim$Hw' is (i) falsity or (ii) the (false) proposition that the book Waverley is a human being. According to the principle of subject matter, the sentence '$\sim$Hw' is in case (i) about falsity (presumably saying that falsity does not hold), and in case (ii) about the proposition mentioned (presumably saying that it does not hold). I believe that the first alternative, which is Frege's result 28-5, would appear to any unprepared reader far less natural than the second.

The foregoing considerations are by no means intended as refutations of or objections to Frege's results. They are merely meant to show that Frege's preliminary explanations of his terms are not sufficient as a basis for his results. In order to understand the specific sense in which Frege means his terms, we have to look not so much at his preliminary explanations as at the reasoning by which he reaches his results. When we do this, we find that Frege makes use of certain assumptions as if they were self-evident or at least familiar and plausible, without formulating them explicitly as the basic principles of his method. These assumptions can be formulated as principles of interchangeability in the following way:

Frege's Principles of Interchangeability

Let $..\mathfrak{A}_j..$ be a complex name containing an occurrence of the name $\mathfrak{A}_j$, and $..\mathfrak{A}_k..$ the corresponding expression with the name $\mathfrak{A}_k$ instead of $\mathfrak{A}_j$.

28-6. *First principle.* If $\mathfrak{A}_j$ and $\mathfrak{A}_k$ have the same nominatum, then $..\mathfrak{A}_j..$ and $..\mathfrak{A}_k..$ have the same nominatum. In other words, the nominatum of the whole expression is a function of the nominata of the names occurring in it.

28-7. *Second principle.* If $\mathfrak{A}_j$ and $\mathfrak{A}_k$ have the same sense, then $..\mathfrak{A}_j..$ and $..\mathfrak{A}_k..$ have the same sense. In other words, the sense of the whole expression is a function of the senses of the names occurring in it.

Now let us see how Frege reaches his results 28-4 and 28-5 with the help of the first principle. His problem is: What is the nominatum and what is the sense of an (isolated) sentence? He says: "If we replace a word in a sentence by another word with the same nominatum but a different sense, then this change cannot have any influence upon the nominatum of the whole sentence."[23] Here, the first principle seems to be tacitly presupposed. Let us take two sentences which are alike except for the occurrence of the phrases 'the morning star' in the one and 'the evening star' in the other. According to our earlier statements (28-1 and 28-3), this is a case in question. Hence, according to Frege's reasoning just quoted, the two sentences have the same nominatum. What, then, could be regarded as this common nominatum? The propositions expressed by the two sentences may, obviously, be different. Hence they cannot be the nominata; therefore, Frege reasons, they must be the senses of the sentences. (Here another assumption seems to be tacitly made, namely, that the proposition expressed by a sentence, because it has clearly a close (semantical) relation to the sentence, must be either its nominatum or its sense.) On the other hand, the two sentences have the same truth-value (at least in ordinary cases). Therefore, the truth-value may be regarded

[23] [Sinn], p. 32.

as the common nominatum. Thus the results 28-4 and 28-5 are reached (for ordinary cases).

The most important application of Frege's two principles is to cases in which the whole expression . . $\mathfrak{A}_j$ . . is an isolated sentence (while $\mathfrak{A}_j$ may be either a sentence or a name of another form). For these cases the principles take the following special forms, if the results 28-4 and 28-5 are applied to the whole sentences:

Frege's Principles of Interchangeability within Sentences

Let . . $\mathfrak{A}_j$ . . be an isolated sentence containing an occurrence of the name $\mathfrak{A}_j$, and . . $\mathfrak{A}_k$ . . the corresponding sentence with the name $\mathfrak{A}_k$ instead of $\mathfrak{A}_j$.

28-8. *First principle.* If $\mathfrak{A}_j$ and $\mathfrak{A}_k$ have the same nominatum, then . . $\mathfrak{A}_j$ . . and . . $\mathfrak{A}_k$ . . have the same truth-value. In our terminology (11-1): Names which have the same nominatum are interchangeable with one another.

28-9. *Second principle.* If $\mathfrak{A}_j$ and $\mathfrak{A}_k$ have the same sense, then . . $\mathfrak{A}_j$ . . and . . $\mathfrak{A}_k$ . . express the same proposition. In our terminology: Names which have the same sense are L-interchangeable with one another.

Our references in what follows are to these specialized forms of Frege's two principles.

What Frege means by 'nominatum' and 'sense' is shown more clearly by these principles than by his preliminary explanations. Frege's first principle 28-8 is the same as 24-3a, the principle of interchangeability for the name-relation. Since Frege's discussion shows that the principles 24-1 and 24-2 also hold for his concept of nominatum, his method is a particular form of what we have called the method of the name-relation. As we have seen earlier, 24-3a is quite plausible; hence Frege's first principle is plausible. Whether this is also true for his second principle is hard to say. But I think it does not seem implausible if we regard it as revealing the fact that Frege understands the term 'sense' in such a way that the sense of a compound expression and, in particular, of a sentence is something which is determined by the senses of the names occurring in it.

Frege's principles lead him, on the one hand, to the results 28-4 and 28-5 for ordinary cases—for example, for isolated sentences—as we have seen. On the other hand, these same principles compel him to regard certain cases as exceptions to these results and thereby to make his whole scheme rather complicated. These exceptions are the cases in which a name occurs in an *oblique* context (which is about the same as a nonextensional context in our terminology, 11-2a). Take, for example, the occurrence of the (false) sentence

(i) 'the planetary orbits are circles'

within the oblique context

(ii) 'Copernicus asserts that the planetary orbits are circles'.

The problems involved here would, of course, be the same if, instead of 'asserts', a term like 'believes' were to occur; hence this example is similar to the belief-sentences discussed earlier (§ 13). According to Frege's results (28-5 and 28-4), the ordinary nominatum of (i), that is, that nominatum which this sentence has when occurring either isolated or in an ordinary, nonoblique context, is its truth-value, which happens to be falsity; and the ordinary sense of (i) is the proposition that the planetary orbits are circles. Now Frege says that the sentence (i) within the oblique context (ii) has not its ordinary nominatum but a different one, which he calls its oblique nominatum, and not its ordinary sense but a different one, which he calls its oblique sense. Concerning the oblique nominatum, Frege makes the following two statements; the second is a special case following from the first:

28-10. The oblique nominatum of a name is the same as its ordinary sense.

28-11. The oblique nominatum of a sentence is not its truth-value but the proposition which is its ordinary sense.

Thus, for the above example the following result holds:

28-12. The oblique nominatum of the sentence (i), that is, the entity named by (i) in an oblique context like (ii), is the proposition that the planetary orbits are circles.

For this result, Frege gives two reasons at different places in his paper. (1) "In the oblique mode of speech, one speaks about the sense, for example, of the utterance of another person. Hence it is clear that . . . in this mode of speech a word does not have its ordinary nominatum, but names that which ordinarily is its sense."[24] I understand Frege's reasoning here in the following way, if applied to the above example. He seems to presuppose tacitly the principle of subject matter (24-2). According to it, the whole sentence (ii) speaks about the nominatum of the subsentence (i). Now it is clear that (ii) does not speak about the sentence (i), because Copernicus may have used other words than (i) and even another lan-

[24] *Ibid.*, p. 28.

guage. Nor does (ii) speak about the truth-value of Copernicus' statement but rather about its sense, because (ii) says that Copernicus asserted a certain sense, a certain proposition, namely, that proposition which is the ordinary sense of (i). Therefore, this proposition must be the nominatum of (i) in (ii). (2) That the nominatum of a sentence in an oblique context is not the truth-value but the proposition, is, Frege says, "also to be seen from [the circumstance] that it is irrelevant for the truth of the whole sentence whether that proposition is true or false."[25] This is presumably meant in the following way: According to Frege's first principle, the nominatum, that is, the truth-value, of the whole sentence (ii) is a function of the nominatum of the subsentence (i). Now if the latter nominatum were the truth-value, then the truth-value of (ii) would depend upon that of (i). This, however, is not the case; in order to establish that (ii) is true we need not know whether (i) is true or false. Hence the nominatum of (i) in (ii) cannot be its truth-value; therefore, it must be the proposition. (For this last step, again, a certain assumption seems tacitly presupposed.)

In one respect, Frege's concept of proposition ('Gedanke') is not quite clear; he does not state an identity condition for propositions. In the foregoing discussion I have assumed that he takes the same identity condition that we take, namely, L-equivalence (see § 6 and [I], p. 92). However, in this case, Frege's analysis of sentences with terms like 'asserts', 'believes', etc., is not quite correct; because a sentence of this kind may change its truth-value and hence, a fortiori, its sense if the subsentence is replaced by an L-equivalent one (see, for example, the discussion of belief-sentences in § 13, especially 13-4). His analysis would be correct if he had in mind a condition stronger than L-equivalence, something similar to the concept of intensional structure explained above (§ 14). In this case our second formulation of 28-9, which was meant as a translation of Frege's second principle into our terminology, must be omitted.

§ 29. Nominatum and Sense: Extension and Intension

Frege's pair of concepts (nominatum and sense) is compared with our pair (extension and intension). The two pairs coincide in ordinary (extensional) contexts, but not in oblique (nonextensional) contexts. This does not constitute an incompatibility, a theoretical difference of opinion, but merely a practical difference of methods. Frege's pair of concepts is intended as an explicatum for a certain traditional distinction, and our pair as an explicatum for another distinction.

[25] *Ibid.*, p. 37.

We shall now compare Frege's distinction between the nominatum and the sense of an expression with our distinction between the extension and the intension of an expression.

Our pair of concepts is, like Frege's, intended to serve for the purposes of semantical meaning analysis. Our two concepts may be regarded, like Frege's, as representing two components of meaning (in a wide sense). The concepts of sense and of intension refer to meaning in a strict sense, as that which is grasped when we understand an expression without knowing the facts; the concepts of nominatum and of extension refer to the application of the expression, depending upon facts.

A decisive difference between our method and Frege's consists in the fact that our concepts, in distinction to Frege's, are independent of the context. An expression in a well-constructed language system always has the same extension and the same intension; but in some contexts it has its ordinary nominatum and its ordinary sense, in other contexts its oblique nominatum and its oblique sense.

Let us, first, compare the extension of an expression with its ordinary nominatum; it seems that these concepts coincide. With respect to predicators, Frege does not seem to have explained how his concepts are to be applied; however, I think that Church[26] is in accord with Frege's intentions when he regards a class as the (ordinary) nominatum of a predicator (of degree one)—for instance, a common noun—and a property as its (ordinary) sense. As an example, Church states that the nominatum of 'unicorn' is the null class, and its sense is the property of unicorn-hood. And here the extension is likewise the class in question. With respect to a sentence, its truth-value is both the ordinary nominatum and the extension. And in the case of an individual expression the ordinary nominatum and the extension is the individual in question. Thus we have this result:

29-1. For any expression, its ordinary nominatum (in Frege's method) is the same as its extension (in our method).

It is more difficult to see clearly what constitutes the ordinary sense in Frege's method. As mentioned before, this is due to the lack of precise explanation and especially of a statement as to the condition of identity of sense; we shall assume here again that Frege would agree to take L-equivalence as this condition. Then, for a sentence, its ordinary sense is the proposition expressed by it, hence it is the same as its intension. For a predicator (of degree one) its ordinary sense is the property in question, and its intension is the same. Frege does not use any special term for the

[26] [Review C.], p. 301.

sense of an individual expression.[27] But he says that the sense of a sentence is not changed if an individual expression occurring in an ordinary context is replaced by another one with the same sense. Therefore, it seems reasonable to assume that what he means by the sense of an individual expression is about the same as what we mean by an individual concept. Hence, on the basis of our understanding of Frege's explanations, the following seems to hold:

29-2. For any expression, its ordinary sense (in Frege's method) is the same as its intension (in our method).

Thus, for ordinary occurrences of expressions, our two concepts coincide with those of Frege. The differences arise only with respect to expressions in an oblique context. Here our concepts lead to the same entities as for the ordinary occurrences of the same expressions, while Frege's concepts lead to different entities. As we have seen earlier, this complication is not introduced by Frege arbitrarily but is an inevitable consequence of his general principles, especially the first.

It seems that Frege, in introducing the distinction between nominatum and sense, had the intention of making more precise a certain distinction which had been made in various forms in traditional logic. Thus his task was one of explication (in the sense explained in the beginning of § 2). The explicata proposed by him are the concepts of nominatum and sense. Now the question is: What were his explicanda, that is, for which pair of traditional concepts did Frege propose his explicata? Church[28] refers in this connection, first, to the distinction between 'extension' and 'comprehension' in the Port-Royal Logic, and, second, to the distinction between 'denotation' and 'connotation' made by John Stuart Mill. It seems to me that we find in the historical development *two* pairs of correlated concepts, appearing in various forms. These pairs are closely related to each other and may sometimes even merge. Nevertheless, I think that it is, in general, possible to distinguish them. (1) In traditional logic we often find two correlated concepts: on the one hand, what was called the 'extension' or 'denotation' (in the sense of J. S. Mill) of a term or a concept; on the other hand, what was called its 'intension', 'comprehension', 'meaning', or 'connotation'.[29] It seems to me that Frege intended an explication of this

[27] Church uses the term 'description', which is, however, more customary for an individual expression constructed with an iota-operator than for its sense.

[28] [Review C.], p. 301.

[29] For a detailed discussion and comparison of the conceptions of Mill and other authors see Ralph M. Eaton, *General Logic* (1931), chap. vi.

pair of concepts by his distinction between the value-distribution of a propositional function and the propositional function itself; in the case of degree one, this distinction is the familiar one between a class and a property. Our distinction between extension and intension is likewise meant as an explication of the same pair of concepts, as far as predicators are concerned, and simultaneously as an enlargement of the domain of application of the customary concepts to other kinds of designators. (2) The second pair of concepts starts with the name-relation. In everyday language, it is said, for instance, that 'Walter Scott' is a name of the man Walter Scott. Logicians extend the application of this relation. They also regard individual descriptions as names, e.g., 'the author of Waverley' as a name of the same man Walter Scott, a usage not admitted by everyday language. Going further, they even construe expressions of another than the individual type as names; they regard them as names of abstract entities, e.g., of classes or properties, relations, functions, propositions, etc. (Other terms used as synonyms of 'is a name of' were mentioned at the beginning of § 24.) With respect to any expression regarded as a name, a distinction is made here between that entity whose name the expression is and the meaning or sense of the expression. It seems that the second concept in this pair is very similar to the second in the first pair; for both of them the term 'meaning' is sometimes used.

Now it seems to me that the explicandum which Frege intended to explicate by his distinction between nominatum and sense was the second pair of concepts rather than the first. And I interpret also some of Quine's discussions in [Notes] as an endeavor toward a clarification and explication of the concepts of the second pair. Since Church's discussions in recent publications, especially [Review C.] and [Review Q.], are intended to defend and develop Frege's distinction, I regard them, too, as belonging more to the second historical line than to the first. However, the two historical lines, the two pairs of concepts taken as explicanda, are closely related to each other. I have emphasized the difference between them only in order to make clearer the difference between the problem which Frege intended to solve and my problem or, more exactly, the difference between the explicandum which Frege took as the basis of his distinction between nominatum and sense (if I understand him correctly) and the explicandum for which my distinction between extension and intension is intended.

Thus it becomes clear—and I wish to emphasize this point—that the difference between Frege's method and that here proposed is not a difference of opinion. In other words, it is not the case that there is one question to which different and incompatible answers have been given. There are two questions, and, more precisely, these are not even theoretical questions but merely practical aims. While the general aim is the same, namely, the construction of a pair of concepts suitable as instruments for semantical analysis, the specific aims are different. Frege tries to achieve the general aim by an explication of one pair of concepts, I by the explication of another pair. Frege's principles are not assertions which are open to refutation or doubt. They are to be regarded rather as part of the characterization of his two concepts and hence hold analytically for these concepts. If someone were to say—as I do not—that he disagrees with Frege's principles, he would merely be saying in effect that he understands the two terms 'nominatum' and 'sense' in a way different from Frege—in other words, that he uses different concepts—and there would be thus no genuine disagreement. The results found by Frege, including the complication in the case of oblique contexts, are consequences of his principles and hence share their analytic validity (assuming that Frege made no mistake in reasoning from the principles to the results). Therefore, I am in complete agreement with Frege's results in this sense: they are valid for his concepts. The same holds for Church's results on the same (or a somewhat modified) basis.

The two concepts used in our method coincide, as we have seen earlier, in certain cases with Frege's concepts, while in other cases they do not. This is not a contradiction between two theories, since our concepts are admittedly different from Frege's. The situation is, rather, similar to the following: Suppose someone divides all animals into aquatic, aerial, and terrestrial animals; someone else divides them into fishes, birds, and the rest. The two classifications coincide to some extent because fishes are aquatic animals and birds are aerial; but they do not coincide entirely. The one man puts whales into his first class, while the other does not. This fact, however, does not constitute a difference of opinion, a theoretical contradiction, because the two concepts in question are admittedly different. Since the two classifications and the assertions made on their bases are not incompatible, it would be theoretically possible to use both simultaneously. However, if the simultaneous use of both seems unnecessarily complicated, there is a kind of practical incompatibility or competition. In this case the decisive question is this: which of the two triples of concepts is more fruitful for the purpose for which both are proposed, namely, a classification of animals?

The situation with regard to Frege's pair of concepts and that proposed here seems to me to be analogous. I have the feeling, without being quite certain, that it would not be very fruitful to use simultaneously both pairs of concepts for semantical analysis. If so, then there is, in spite of the theoretical compatibility, a practical competition or conflict. This conflict might, for instance, appear over the following point, which has been mentioned earlier: A logician, thinking in terms of Frege's concepts, might be inclined, though not compelled, to construct a logical system in such a way that it contains different expressions for classes and for properties, while a logician, thinking, instead, in terms of extension and intension, would probably be less inclined to do so.

§ 30. The Disadvantages of Frege's Method

Frege's special form of the method of the name-relation involves additional complications. Starting with any ordinary name, it leads to an infinite number of entities and an infinite number of expressions as names for them, while the method of extension and intension needs only one expression and speaks only of two entities. Furthermore, according to Frege's method, the same name, when occurring in different contexts, may have an infinite number of different nominata; and sometimes even the same occurrence of a name may simultaneously have several nominata.

The disadvantages of Frege's pair of concepts in comparison with the pair here proposed all belong to the concept of nominatum. Frege's concept of sense is very similar to that of intension; we might even say that, when we consider simply these two concepts, it is difficult to see any reason that there should be a difference between them. The difference is brought about by Frege's differentiation between the ordinary and the oblique sense of a name. It is not easy to say what his reasons were for regarding them as different. Perhaps he was led to make this distinction because of his original distinction between the ordinary and the oblique nominatum. It does not appear, at least not to me, that it would be unnatural or implausible to ascribe its ordinary sense to a name in an oblique context. However, Frege could not do this because he had already used this ordinary sense as nominatum in the oblique context. And since he assumes that nominatum and sense must always be different, he had thus to introduce a third entity as the oblique sense. Incidentally, it seems that Frege nowhere explains in more customary terms what this third entity is.

Since Frege's method is a special form of what we have called the method of the name-relation, it also possesses the disadvantages which we have previously found in this method. We found (§ 25) that the concept of nominatum involves a certain ambiguity, which is also transferred to other semantical concepts, for instance, those of identity sentence and identity sign.

Further, we saw (§§ 26, 27) that the method of the name-relation may lead to a complicated duplication or multiplicity of names within the same type. If Frege's form of the method is adopted, the situation becomes even more complicated. We shall illustrate this by two examples. (See the diagram, where an arrow with 'N' indicates the name-relation and an arrow with 'S' the sense-relation.) *Example (1)*: Let us start with a name

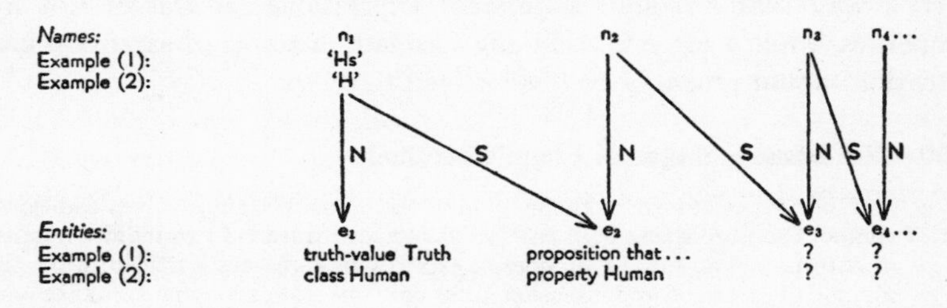

n_1, say the sentence 'Hs'. According to Frege's method, there is an entity, e_1, named by this name; this is the truth-value of 'Hs'. And there is another entity, e_2, which is the sense of 'Hs'; this is the proposition that Scott is human. This proposition e_2 may also have a name; if we wish to speak about it, we need a name for it. This name is different from n_1 because the latter is the name of e_1 and hence, in a well-constructed language, should not be used simultaneously as a name of another entity. Let the new name be n_2. Like any name, n_2 has a sense. This sense of n_2 must be different from the nominatum of n_2; it is a new entity, e_3, not occurring in customary analyses. In order to speak about e_3, we need a new name, n_3. The sense of n_3 is a new entity e_4; and so forth *ad infinitum*. *Example (2)*: The situation is analogous if the first name n_1 is of another type, for instance, a predicator, say 'H'. The entity e_1 named by n_1 is here the class Human; the sense e_2 is the property Human. The name n_2 is introduced as a name for the property Human; and the new entity e_3 is the sense of this name. The name n_3 is a name of this sense e_3; e_4 is the sense of this name n_3, and so on. Generally speaking, if we start with any name of a customary form, we have, first, two entities familiar to us: its ordinary nominatum and its ordinary sense; they are the same as its extension and its intension, respectively. Then Frege's method leads, further, to an infinite number of entities of new and unfamiliar kinds; and, if we wish to be able to speak about all of them, the language must contain an infinite number of names for these entities. To provide for this infinite sequence of names seems, thus, a natural decision on the basis of Frege's method. And Church does, indeed, take this decision in his development of Frege's method by declaring it desirable "that the object language should contain for every name in it a name of the associated sense."[30] On the basis of the method of extension and intension, on the other hand, we need in the object language, instead of an infinite sequence of expressions, only one expression (for instance, in the first example 'Hs', in the second 'H'); and we speak in the metalanguage only of two entities in connection with the one expression, namely, its extension and its intension (and even these are, as we shall see later, merely alternative ways of saying the same thing).

The fact that, according to Frege's method, the same name may have different nominata in different contexts has already been mentioned as a disadvantage. But the multiplication of entities goes far beyond Frege's initial distinction between the ordinary and the oblique nominatum of a name. Actually, these two nominata constitute only the beginning of an infinite sequence of nominata for the same name. If we apply Frege's method to sentences with multiple obliqueness, then we have to distinguish the ordinary nominatum of the name, its first oblique nominatum, its second oblique nominatum, and so forth. In order to construct an example, let us suppose that the system *S* contains not only, like S_2 (see § 11, Example II), modal signs, say 'Np' for 'it is necessary that p' and '$\Diamond p$' for 'it is possible that p', but also psychological terms, say 'Jp' for 'John believes that p'. Now let us consider a series of sentences in *S*, each occurring within the next in a simple oblique context:

(i) 'Hs' ('Scott is human');
(ii) '$\Diamond$(Hs)' ('it is possible that Scott is human');
(iii) 'J($\Diamond$(Hs))' ('John believes that it is possible that Scott is human');
(iv) '$\sim$N(J($\Diamond$(Hs)))' ('it is not necessary that John believes that it is possible that Scott is human'); etc.

Let us see what the nominatum of the original sentence 'Hs' is in these various contexts. According to our previous explanation of Frege's method, the nominatum of 'Hs' in isolation is its truth-value, hence the entity e_1 in the above diagram; and the nominatum of its occurrence within (ii) is the proposition that Scott is human, hence the entity e_2 in the diagram. It can further be shown, by an analysis which we shall not describe here in detail, that the nominatum of 'Hs' within (iii) is e_3, its nominatum within (iv) is e_4, and so on. Thus the same expression 'Hs' has an infinite number of different entities as nominata when it occurs in different contexts.

[30] [Review Q.], p. 47.

This fact—that different occurrences of a name may have different nominata—is certainly a disadvantage. It is the reason that Church proposes a certain modification of Frege's method whereby this multiplicity of nominata is avoided (see § 32, Method III).

Worse than the multiplicity of nominata for different occurrences of a name is the fact that within certain contexts, according to Frege's own analysis, one occurrence of a sentence has simultaneously two different nominata. Frege takes as an example a sentence 'Bebel wähnt, dass . . .', that is (writing 'A' as an abbreviation for a long subsentence), 'Bebel has the illusion that A', or 'Bebel believes erroneously that A'. Frege interprets this sentence, no doubt correctly, as 'Bebel believes that A; and not A'. Now here we have two occurrences of 'A', the first in an oblique context, the second in an ordinary one, with therefore different nominata. Thus Frege comes to the conclusion that, in the original sentence 'Bebel believes erroneously that A', the subsentence 'A' "strictly speaking, must be taken twice with different nominata of which the one is a proposition, the other a truth-value".[31] The situation is analogous in a case like 'John knows that A', because this implies 'John believes that A; and A'.

This double nominatum of a name, not, as in the earlier cases, for different occurrences but for the same occurrence, seems a startling result of Frege's method. The sentences in question seem perfectly clear. At first glance it will not seem plausible that the subsentence 'A' should simultaneously name two distinct entities. It can easily be seen that the feature here discussed has nothing to do with the ordinary ambiguities so frequently met with in natural word languages, but is likewise to be found in an exact, symbolic system of modal logic. A modal sign 'CT' for contingent truth of propositions (which is a nonsemantical concept, see § 23) can be introduced in S_2 on the basis of 'N' (see § 11, Example II) in this way:

30-1. *Abbreviation.* 'CT(p)' for '$p \bullet \sim$N(p)'.

On this basis, the sentence 'CT(Hs)' is L-equivalent to 'Hs $\bullet \sim$ N(Hs)'; in words: 'Scott is human, but it is not necessary that Scott is human'; or, briefly: 'Scott happens to be human' According to Frege's analysis, the sentence 'Hs' within 'CT(Hs)' has at once two different nominata, as have the signs 'H' and 's'; and the same holds for the words 'Scott' and 'human' in the sentence 'Scott happens to be human'. This seems a rather unsatisfactory result.

If, instead of Frege's method, the method of extension and intension is

[31] [Sinn], p. 48.

used, then the situation becomes much simpler. Every expression has always the same extension and the same intension, independent of the context. The problems connected with modal contexts will be discussed later (chap. v).

§ 31. The Antinomy of the Name-Relation

> The third principle of the name-relation (24-3) permits replacing a name with another name of the same entity. Although this principle seems quite plausible, it is not always valid. This has been pointed out by Frege, Russell, and Quine. The contradiction which sometimes arises if such a replacement is made in a nonextensional context is called here the antinomy of the name-relation.

The principles which characterize the method of the name-relation (24-1, 2, and 3) seem quite plausible; and this holds for either form of the principle of interchangeability, the one using the concept of name-relation (24-3a) and the other using the concept of identity (24-3b). Therefore, in a naïve approach without a closer investigation, we might be tempted to regard these principles as generally valid without any restrictions. However, if we do so and, in particular, if we apply the principle of interchangeability in either form to nonextensional contexts, we arrive at a contradiction. I propose to call this contradiction the *antinomy of the name-relation*. [My choice of this term is, of course, motivated by the fact that, from my point of view, the method of the name-relation is responsible for the antinomy. Others, who regard this method as harmless and unobjectionable and who feel that the source of the difficulty lies, rather, in the use of modal contexts or, more generally, intensional contexts or, still more generally, oblique (i.e., nonextensional) contexts, will perhaps prefer to call it the antinomy of modality or of intensionality or of obliquity.]

The antinomy of the name-relation can be constructed, as we shall see, in either of two forms; the first uses the first form of the principle of interchangeability (24-3a), the second uses its second form (24-3b). The second form of the antinomy may perhaps also be called *antinomy of identity* or antinomy of identical nominata or *antinomy of synonymity* (provided the term 'synonymous' is understood, not in the sense of 'intensionally isomorphic' (14-1), but as 'having the same nominatum').

Frege was the first to point out the circumstance that the principle of interchangeability (see 24-5) if applied to the ordinary nominata of names does not hold for oblique contexts. Although Frege's formulation was not presented in terms of a contradiction, his result constitutes the basis of what I propose to call the antinomy of the name-relation.

It seems that the antinomic, paradoxical character of the situation was first seen by Russell.[32] He explains the antinomy in its second form with respect to an interchange of individual expressions as the first of the three "puzzles" which he says every theory of denoting (name-relation) must solve. He states the second form of the principle of interchangeability (24-3b) in the following words: "If a is identical with b, whatever is true of the one is true of the other, and either may be substituted for the other in any proposition without altering the truth or falsehood of that proposition."[33] He takes as an example the sentence 'George IV wished to know whether Scott was the author of Waverley'. If in this sentence, on the basis of the true identity sentence 'the author of Waverley is identical with Scott' (9-1), the description 'the author of Waverley' is replaced by 'Scott', the resulting sentence is presumably false.

Quine[34] likewise points out the second form of the antinomy with respect to individual expressions. His first examples are psychological sentences with the phrases 'is unaware that' and 'believes that';[35] they are similar to Frege's example, 'Copernicus asserts that . . .' (see above, § 28), and Russell's example just mentioned. Further examples given by Quine are modal sentences.[36] The first is: 'Necessarily, if there is life on the evening star, then there is life on the evening star'. If here, on the basis of the identity sentence, 'The morning star is the same as the evening star' (28-2), which is found to be true by astronomical observations, one occurrence of 'the evening star' is replaced by 'the morning star', a false sentence results. (If, instead of the truth of the identity sentence 28-2, the semantical statement 28-1 is used, we have the first form of the antinomy.) In another example of a modal sentence, Quine uses numerical expressions:

'9 is necessarily greater than 7'.

If here, on the basis of the true identity sentence 'The number of planets = 9', '9' is replaced by 'the number of planets', the following false sentence results:

'The number of planets is necessarily greater than 7'.

I shall now give an example of the antinomy in both forms with respect to predicators. We found earlier an ambiguity in the concept of the nominatum of a predicator (for example, the German word 'gross' may be re-

[32] [Denoting], p. 485.

[33] *Ibid.*

[34] [Notes], p. 115.

[35] *Ibid.*

[36] *Ibid.*, p. 121.

garded as a name of the class Large or of the property Large, see § 25). In order to show that the antinomy of the name-relation is independent of this ambiguity, the example will be formulated with phrases of the form 'the class . . .' and only classes taken as nominata of these phrases. The following sentence is true ('necessary' is here, as in earlier examples, used in the sense of 'logically necessary'):

> 'It is necessary that the class Featherless Biped is a subclass of the class Biped'.

Now we replace in this sentence 'the class Featherless Biped' by 'the class Human'; this replacement may be based either, according to 24-3b, on the circumstance that the identity sentence 'the class Featherless Biped is the same as the class Human' is true (4-7) or, according to 24-3a, on the circumstance that the phrases 'the class Featherless Biped' and 'the class Human' have the same nominatum. The result of the replacement is the sentence

> 'It is necessary that the class Human is a subclass of the class Biped'.

Since, however, the fact that human beings have two legs is a contingent biological fact and not logically necessary, the following is true:

> 'It is not necessary that the class Human is a subclass of the class Biped'.

The contradiction between these two results constitutes an instance of the antinomy of the name-relation.

Those logical situations which are called logical antinomies (in the modern, not the Kantian sense) or logical paradoxes are characterized by the fact that there are two methods of reasoning, which, although both plausible and in accordance with customary ways of thinking, lead to contradictory conclusions. Any solution of an antinomy, that is, the elimination of the contradiction, consists, therefore, in making suitable changes in the reasoning procedure; at least one of its assumptions or rules must, in spite of its plausibility, be abolished or restricted in such a way that it is no longer possible to reach the two incompatible conclusions. Sometimes a certain form of inference is abolished or restricted. Sometimes a more radical step is taken by abandoning certain forms of sentences which were previously regarded as meaningful and harmless. Thus, for instance, Russell's solution of the antinomy known by his name consisted in the rejection of sentences of the form '$\alpha\epsilon\alpha$'. Sometimes several different ways for solving a given antinomy are found. It is a matter of theoretical investigation to discover the consequences to which each of the solutions leads and, especially, what sacrifices of customary and plausible ways of expression or deduction each of them entails. But which of the solutions we choose for the construction of a language system is ultimately a matter of practical decision, influenced, of course, by the results of the theoretical investigation.

§ 32. Solutions of the Antinomy

Six procedures for the solution of the antinomy of the name-relation are dis cussed. The first five still apply the method of the name-relation. Frege and Church develop particular forms of this method by introducing certain distinctions, which, however, lead to a more complicated language. Russell restricts to a considerable degree the application of the method of the name-relation and thereby of the semantical analysis of the meaning of expressions. Quine does the same to a smaller degree. The antinomy would also be eliminated by restricting the language to extensional sentences; but it is not known at present whether the whole of logic and science is expressible in a language of this kind. Finally, the method of extension and intension avoids the antinomy by avoiding the concept of nominatum. The concept of extension, though similar to that of nominatum, eliminates the contradiction without unnatural restrictions and complications.

We shall now explain some of the solutions for the antinomy of the name-relation which have been proposed or considered by logicians; we call them Methods I–VI. First, we discuss five solutions which preserve the method of the name-relation, at least to some extent. They may be regarded as particular forms of this method. We shall find that each of them has serious disadvantages. Then we shall consider the possibility of solving the antinomy by giving up the method of the name-relation.

Method I, *Frege*. It seems that Frege was aware of the fact that the principle of interchangeability (in the form 24-3a) would lead to a contradiction if the ordinary nominata of names were ascribed also to their oblique occurrences and that the contradiction does not arise if different nominata are ascribed to these occurrences. In this sense we may say that Frege offers a solution for the antinomy of the name-relation. It is true that Frege does not speak explicitly of the necessity of avoiding a contradiction; he gives other reasons for his distinction between the ordinary nominatum and the oblique nominatum of a name. His reasoning gives the impression that this distinction appeared to him natural in itself, without regard to any possible contradiction. However, I think that to many readers it will scarcely appear very natural and that they, like myself, will see the strongest argument in favor of Frege's method rather in the fact that it is a way of solving the antinomy.

The disadvantages of Frege's method have been explained earlier (§ 30). We have seen that the unnecessary multiplicity of entities and names which is generally a consequence of the method of the name-relation is here even much greater. Furthermore, occurrences of the same name may have different nominata—indeed, an infinite number of them; and in certain contexts even the same occurrence of a name may have simultaneously several nominata.

Method II, *Quine*. Quine[37] uses the term 'designation' for the name-relation. He says of an occurrence of an expression in a nonextensional context (as, for instance, 'the evening star' in the first and '9' in the second of the two examples of his, quoted in the preceding section) that it is "not purely designative" and that it does not refer simply to the object designated (the nominatum). He thinks that nonextensional contexts are fundamentally different from extensional contexts and more similar to contexts in quotation marks; and, in particular, that the customary logical rules of specification and existential generalization are not valid for nonextensional contexts (this will be discussed later, § 44). Thus his solution agrees with Frege's in not ascribing the ordinary nominatum to an occurrence of a name in a nonextensional context. But where Frege ascribes a different nominatum, Quine ascribes no nominatum at all. Consequently, the principle of interchangeability (see his formulation 24-6) is declared by Quine not to be applicable to these occurrences, and thus the antinomy is eliminated.

The advantage of Quine's method in comparison with Frege's consists in avoiding the immense multiplication of entities and corresponding names to which the latter method leads. But Quine's method pays a high price for this simplification by restricting the name-relation ('designation') to extensional contexts and grouping all nonextensional contexts together with contexts in quotation marks and, further, by imposing narrow restrictions upon the use of variables in modal sentences. Those logicians in particular who are interested in constructing or in semantically analyzing systems of modal logic will hardly be inclined to adopt this method.

Method III, *Church*. Church[38] regards Frege's method as preferable to Quine's in two respects: first, because it provides that a name always has a nominatum[39] even in nonextensional contexts and, second, because Frege's conception of the sense of names as something outside the language (e.g., propositions or properties) seems more natural than Quine's way of construing the sense (meaning) of a name as its L-equiva-

[37] [Notes]. [38] [Review Q.]. [39] *Ibid.*, p. 46.

lence class (see end of § 33). However, Church does not simply adopt Frege's method in its original form; he proposes important modifications in it. He agrees with Frege's conclusion that the nominatum of an oblique (nonextensional) occurrence of a name must be different from its ordinary nominatum and must be the same as its ordinary sense. But Church seems to accept this only as a result of an analysis of nonextensional sentences as they occur in natural word languages and in systems of modal logic of the customary form. In a well-constructed language, however, this multiplicity of nominata for the same name should be avoided. Therefore, Church proposes, for semantical discussions in the natural word languages, "to adopt some notational device to distinguish the oblique use of a name from its ordinary use";[40] this would be analogous to the customary use of quotation marks. Mere distinguishing marks are not sufficient, however, in a symbolic language system; here we should go one step further, as we do when we use not quotation marks but special symbols as names of signs. "In a formalized logical system, a name would be represented by a distinct symbol in its ordinary and in its oblique use".

I agree that, if the method of the name-relation is used, then the changes in the notation proposed by Church are indeed an improvement. On the other hand, it seems that these changes would cause an additional complication in a system of modal logic. For example, there would be an infinite number of types corresponding to the one type of sentences in the method of extension and intension.

Although Church's method avoids the multiplicity of nominata for the same name, it shares the other complications of the original form of Frege's method explained in § 30. This fact, however, is not an argument against Church's method in comparison to the other forms of the method of the name-relation. On the contrary, I think that Church's form of the method may well be regarded as that which carries out the basic ideas of the method of the name-relation in the most consistent and thorough way, eliminating features not tolerable in a well-constructed system and not restricting unduly the domain of application of the fundamental concepts of the method. Therefore, the great complications to which it leads are to be regarded, rather, as an argument against the method of the name-relation in general—provided that there is some other convenient method which avoids them.

Method IV, *Russell*. Russell[41] constructs the antinomy of the name-relation with respect to individual expressions; in his example (see the

[40] *Ibid.*, p. 46.

[41] [Denoting].

preceding section) the description 'the author of Waverley' is replaced by the proper name 'Scott'. According to Russell's conception, a description has no meaning in itself, but a sentence containing a description has a meaning,[42] and this meaning can be expressed without using the description. The contextual definition of a description (see above, § 7, Method II) is a rule for transforming a sentence containing a description into a sentence with the same meaning which no longer contains the description. Although in the case of an individual description which fulfils the uniqueness condition we may regard the one individual (the descriptum) as the nominatum of the description, nevertheless, a sentence containing this description is not about this individual. (Thus the principle of subject matter, 24-2, is rejected with respect to descriptions.) What the sentence actually means is shown only in its expanded form. Proper names (e.g., 'Walter Scott') are regarded as abbreviations of descriptions. Thus, in the primitive notation, neither proper names nor descriptions occur. Therefore, the principle of interchangeability for individual expressions is not applicable, and that form of the antinomy which arises from an interchange of individual expressions is eliminated. The situation is quite analogous for abstraction expressions of classes (for example, '$\hat{x}(\mathrm{H}x)$'; see the explanations above, at the beginning of § 26). These expressions are likewise introduced by contextual definitions and not regarded as having any meaning in themselves. The meaning of a sentence containing a class expression is shown by its expansion in primitive notation, where no class expression occurs. Thus, also with respect to class expressions the principle of interchangeability is inapplicable, and the antinomy does not appear.

If Russell regards sentences as names at all, then presumably he regards them as names of propositions; in any case, he does not regard them as names of truth-values. Thus the final result with respect to Russell's application of the name-relation may be summed up in the following way: Although individual expressions and class expressions may, in a certain sense, be regarded as naming individuals or classes, they do not occur in the primitive notation but are incomplete symbols without independent meaning. As nominata in the strict sense, neither individuals nor classes nor truth-values occur, hence none of those entities which we call extensions. The antinomy of the name-relation arises from an interchange of two expressions with the same nominatum. In all the chief kinds of instances of the antinomy—including all instances mentioned in this book and all instances given by the authors mentioned—the common nomina-

[42] *Ibid.*, p. 480.

tum is an extension. Therefore, Russell's method, by excluding extensions from the realm of nominata in the strict sense, eliminates at least the most important instances of the antinomy.

A few remarks may be made on Russell's objections to Frege's method. The chief objection[43] concerns the case of a description which does not fulfil the uniqueness condition. Frege says that in this case the description has a sense but no nominatum. Russell regards it as unsatisfactory that expressions of the same syntactical form should in one case have a nominatum and in another case not. Since, according to Frege, a sentence is about the nominata of the expressions occurring in it (24-2), in the case in which the uniqueness condition is not fulfilled the sentence is about no entity at all; hence, Russell says,[44] one would suppose that the sentence "ought to be nonsense; but it is not nonsense, since it is plainly false". This reasoning seems to me convincing; moreover, I suppose that Frege himself would agree with it because he regards the feature mentioned as a defect of natural languages.[45] This is the reason for his demand that in a well-constructed language every description should have a nominatum by virtue of a suitable convention.[46] Russell's objection here is that this procedure is artificial and does not give an exact analysis of the actual use of descriptions. However, Frege's convention had a different purpose. He first gave an analysis of the natural language and then proposed the convention as a step not in the exact reconstruction of the natural language but rather in the construction of a new language system intended to be technically superior to the natural language.

Russell's general objections[47] against Frege's distinction between nominatum and sense are rather obscure. This is due chiefly to Russell's confusion between use and mention of expressions, which has already been criticized by Church.[48]

The disadvantage of Russell's method lies in the fact that meaning is denied to individual expressions and class expressions. That these kinds of expressions can be introduced by contextual definitions and hence that what is said with their help can also be said without them is certainly a result of greatest importance but does not seem a sufficient justification for excluding these expressions from the domain of semantical meaning analysis. It must be admitted, I think, that descriptions and class expressions do not possess a meaning of the highest degree of independence; but that holds also for all other kinds of expressions except sentences (see re-

[43] *Ibid.*, pp. 483 f.

[44] *Ibid.*, p. 484.

[45] [Sinn], p. 40.

[46] *Ibid.*, p. 41; see above, § 8.

[47] *Op. cit.*, pp. 485–88.

[48] [Review C.], p. 302.

marks at the end of § 1). And it is certainly useful for the semantical analysis of the meanings of sentences to apply that analysis also to the meanings, however derivative, of the other expressions, in order to show how out of them the independent meanings of the sentences are constituted.

Method V, *Extensional Language*. The most radical method for eliminating any antinomy arising in connection with certain forms of expression consists in excluding these forms entirely. In the case of the antinomy of the name-relation this solution would consist in excluding all nonextensional contexts—in other words, in using a purely extensional language (see the definition 11-2c). To construct an extensional language system for certain restricted purposes involves, of course, no difficulties (as examples of such systems, see, e.g., Quine's language system ML and my systems I and II in [Syntax]). But this is not sufficient for the present purpose. In order to eliminate the antinomy by excluding all nonextensional contexts, it would be necessary to show that for the purposes of any logical or empirical field of investigation an extensional language system can be constructed; in other words, that for any nonextensional system there is an extensional system into which the former can be translated. The assertion to this effect is known as the *thesis of extensionality*.[49] The problem of whether it holds or not is still unsolved. Translatability into extensional sentences has been shown for certain kinds of nonextensional sentences. Thus, for instance, any simple modal sentence is L-equivalent to a semantical sentence in an extensional metalanguage using L-terms, as we shall see later (§ 39).[50] For example, the modal sentence 'N(A)', in words: 'it is necessary that A', is L-equivalent to the semantical sentence ' 'A' is L-true' (according to a convention to be discussed later). The application of this method of translation to sentences with iterated modalities (e.g., 'it is necessary that it is possible that . . .') involves a certain difficulty; this, however, can be overcome, as I have shown at another place.[51] The translation of nonextensional sentences with psychological terms like 'believes', 'knows', etc., is presumably likewise possible, although at present it is not yet clear how it can best be

[49] See [Syntax], § 67; [I], p. 249; Russell, [Inquiry], chap. xix.

[50] For this translation see [Syntax], § 69; I would now define the L-concepts not as syntactical but as semantical concepts (see above, § 2). Note that in this translation the two sentences, although L-equivalent, are not intensionally isomorphic (§ 14). A translation in the stronger sense, preserving intensional structure, is obviously impossible between a nonextensional and an extensional sentence.

[51] [Modalities].

made (see the discussions in §§ 13 and 15). The question of whether an extensional language is sufficient for the purposes of semantics will be discussed later (§ 38); an affirmative answer does not seem implausible, but the question is not yet definitely settled.

If we could prove the thesis of extensionality and if we decided to exclude all nonextensional sentence forms, then obviously the antinomy of the name-relation would be eliminated. Furthermore, the difference between the method of the name-relation and the method of extension and intension would disappear, since, with respect to extensional occurrences, the nominatum of an expression is the same as its extension, and its sense the same as its intension (29-1 and 2). Attractive though these consequences may appear, it seems to me that it would be at least premature to propose Method V as a solution of the antinomy at the present time. Even if the thesis of extensionality were proved, this would not be sufficient as a justification for Method V. We should have to show, in addition, that an extensional language for the whole of logic and science is not only possible but also technically more efficient than nonextensional forms of language. Though extensional sentences follow simpler rules of deduction than nonextensional ones, a nonextensional language often supplies simpler forms of expression; consequently, even the deductive manipulation of a nonextensional sentence is often simpler than that of the complicated extensional sentence into which it would be translated. Thus both forms of language have their advantages; and the problem of where the greater over-all simplicity and efficiency is to be found is still in the balance. Much more investigation of nonextensional, and especially of modal, language systems will have to be done before this problem can be decided. Therefore, for the time being, Method V as a solution of the antinomy has to be left aside.

Method VI, *Extension and Intension.* If, instead of the method of the name-relation, the method of extension and intension is used for semantical analysis, then the concept of nominatum does not occur, and hence the antinomy of the name-relation in its original form cannot arise. Since, however, the concept of extension is in many respects similar to and partly coincides with the concept of nominatum, there might arise, under certain conditions, an antinomy of the identity of extension analogous to that of the identity of nominatum. The antinomy would arise if for the concept of extension a principle analogous to the principle of interchangeability of names (24-3) were laid down. The form which we have chosen for the method of extension and intension excludes the antinomy by prescribing for expressions with the same extension, in other words, for equivalent expressions, a principle of interchangeability which is restricted to extensional contexts (12-1). Our second principle (12-2) concerns L-equivalent expressions, hence those with the same intension; thus it is related to Frege's second principle (28-9).

Perhaps it will occur to the reader at this point to ask why, if a restriction of interchangeability to extensional contexts assures the elimination of the antinomy, we might not simply keep Frege's two concepts and restrict his first principle to extensional (nonoblique) contexts. The reply is that Frege's concept ('bezeichnen') is meant in the sense of a name-relation, that is, as a relation characterized by the principles 24-1 and 2; therefore it would be quite implausible and unnatural, as we have seen earlier, not to maintain the principle of interchangeability 24-3 in its unrestricted form. Or, to put it the other way round, if somebody uses a concept for which the principle 24-3 does not hold unrestrictedly, then this concept is not a name-relation and is not the concept meant by Frege and many other logicians, for example, Church and Quine.

It is easy to see that the method of extension and intension avoids those features of the other methods which we have found to be disadvantages. In our general discussion of the method of the name-relation, we have first explained the ambiguity in the concept of nominatum (§ 25); for instance, even if we understand clearly what is meant by a given predicator, we may regard either the property or the class as its nominatum. The concept of extension does not involve any analogous ambiguity; the extension of any predicator of level one and degree one is the class of those individuals to which the predicator can be truly applied. Further, we have shown the multiplicity of expressions in the object language to which the method of the name-relation leads (§ 26); we have seen that, if our method is used, this multiplicity is replaced by one expression. Further, the complications caused by the particular form of the method introduced by Frege have been explained (§ 30). Their common root is the fact that different occurrences of the same expression may have different nominata. Since the extension of an expression is always the same, independent of the context, no analogous complications are caused by our method. The disadvantage of Quine's method is the restriction of the name-relation to extensional contexts; there is no analogous restriction of the application of the concept of extension. While Church's method avoids some of the disadvantages of Frege's original method, it shares most of them; further, his modification of Frege's method, necessary though it is, causes a new complication, which does not occur in our method. The disadvantage of Russell's method is its denial of meaning to individual expressions and class expressions. In our method there is no such restriction; to every expression of these kinds an extension and an intension are ascribed (for class expressions in the system PM, see above, § 26).

Let us sum up the result of our discussion of the method of the name-relation in this chapter. The method appears in various forms with different authors. Most authors who use the concept of the name-relation do not seem to be aware of the antinomy and do not develop the method in a sufficiently explicit form to enable us to see whether and how they avoid the contradiction. All procedures that have been proposed for the elimination of the antinomy have serious disadvantages; some of these procedures lead to great complications, others restrict considerably the field of application of the semantical meaning analysis. Thus it seems doubtful whether the method of the name-relation is a suitable method for semantical analysis.

CHAPTER IV

ON METALANGUAGES FOR SEMANTICS

In the metalanguage M, which we have used so far, we have spoken about extensions and intensions, for instance, about classes and properties. It is the main purpose of this chapter to show that this distinction does not actually presuppose two kinds of entities but is merely a distinction between two ways of speaking. First, we discuss possible methods for defining extensions in terms of intensions or vice versa, without adopting any of them (§ 33). Then we construct a new metalanguage M′ (§§ 34–36). While M contains distinct expressions for an extension (e.g., 'the class Human') and an intension (e.g., 'the property Human'), M′ contains only one expression (e.g., 'Human'), which is, so to speak, neutral, like the expressions in the symbolic system S_1 (e.g., 'H'). Therefore, we call M′ a *neutral metalanguage*. By this elimination of the duplication of expressions, the apparent duplication of entities disappears. It is shown that all sentences of M can be translated into M′, including the semantics of systems like S_1 (§ 37). Finally, the question is examined as to whether a complete semantical description of a system, even a nonextensional system like S_2, can be formulated in a metalanguage which, in distinction to M and M′, is extensional; it seems that this is the case (§ 38).

§ 33. The Problem of a Reduction of the Entities

In the metalanguage M we have so far spoken as if there were two kinds of entities in each type, extensions and intensions, for example, classes and properties. Here the question is discussed as to whether we can get rid of this apparent duplication of entities by defining one kind in terms of the other. Four methods for defining extensions in terms of intensions are discussed: the conception of extensions as L-determinate intensions (§ 23); Russell's contextual definition of classes in terms of properties, which is shown to involve a certain difficulty; a modified version of Russell's definition, which avoids the difficulty; and, finally, a method which uses property expressions themselves as class expressions but presupposes a particular structure of the language. It does not seem possible to define intensions themselves in terms of extensions. However, the class of all designators L-equivalent to a given designator might be taken as a representative of its intension.

We have used as metalanguage M a part of English, modified and supplemented in a certain way (§ 1). Throughout our discussions we have used in M terms like 'class', 'property', 'truth-value', 'proposition', 'individual', 'individual concept', and the more general terms 'extension' and 'intension'. This manner of speaking gives the appearance of dealing with a great variety of entities and, in particular, with two kinds of entities within each type. As stated at the beginning (§ 4), we have used the terms mentioned only because they help to facilitate understanding, but our theory is not based on the assumption that there are entities of all these kinds. Now, mindful of Occam's razor, we shall try to show how the number of apparent entities can be reduced to half. Since the apparent duplication of entities was actually only a duplication of terminology, all we have to do is to construct another way of speaking which avoids the terminological split into extensions and intensions.

Let us begin with the discussion of predicators, because here the distinction between extension and intension is customary and familiar. If we wish to have a language which is not, like S_1, restricted to elementary statements about things but contains a more comprehensive system of logic and especially of mathematics, then we must introduce means for speaking in general terms not only about things but also about entities of higher levels, say classes or properties. So much is admitted even by those logicians who are most wary in admitting abstract entities.[1] The question is whether it is necessary to admit both kinds of entities, classes and properties, or whether those of the one kind are definable with the help of those of the other. For instance, is one of the two phrases (in M) 'the class Human' and 'the property Human' definable with the help of the other? Explicit definition is not necessary; a contextual definition would suffice to make one of the two phrases dispensable in the primitive formulation.

Let us first look for methods which define class expressions in terms of property expressions.

1. If the concept of L-determinate intensions (§ 22) is available, we can define 'the class *f*' as 'the L-determinate property which is equivalent to the property *f*' (§ 23).

2. If we do not wish to make use of the concept of L-determinate intensions, we may consider the possibility of a contextual definition for 'the class *f*' by a generalized reference to the properties which are equivalent to the property *f*. Since all these properties determine the same class, the most natural procedure seems to be to interpret a statement about the class *f* as a statement about all these properties. Thus, for a system *S*, containing predicate variables '*f*', '*g*', etc., we could lay down the following contextual definition for the class-expression '$\hat{x}(fx)$':

33-1. '. . $\hat{x}(fx)$ . .' for '$(g)[(g \equiv f) \supset \, . \, . \, g \, . \, .]$'.

This definition must be supplemented by a rule specifying what is to be taken, in any given case in which '$\hat{x}(fx)$' occurs, as the context '. . $\hat{x}(fx)$ . .'

[1] See, for example, Quine, [Notes], p. 125: "Anyone who cares to explore the foundations of mathematics must, whatever his private ontological dogma, begin with a provisional tolerance of classes or attributes [i.e., properties]".

to which the definition is to be applied. Following Quine[2] rather than Russell (see below), we stipulate that the definition is to be applied to the smallest sentence or matrix *in the primitive notation* in which the class expression occurs. Thus, before applying the definition, we have to transform the given sentence containing a class expression by eliminating all previously defined signs with the help of their definitions; then, with the help of definition 33-1, we expand each smallest matrix in which the class expression occurs.

3. Russell[3] was the first to propose a contextual definition of class expressions on the basis of property expressions. Whitehead and Russell used this definition in their construction of the system of mathematics in [P.M.].[4] Though the method has been able to supply a good working basis for this construction, there is one feature of the definition which seems to me disadvantageous. The definition given above (33-1) is, indeed, nothing else than a variant of Russell's definition, changed, however, with respect to the point in question. The definition in [P.M.], transcribed in our notation,[5] is as follows:

33-2. '$\ldots \hat{z}(fz) \ldots$' for '$(\exists g)[(g \equiv f) \bullet \ldots g \ldots]$'.

The definiens here contains an existential quantifier, not a universal one, as does 33-1. Thus a statement about the class f is here interpreted as a statement not about all properties but about at least one property equivalent to the property f (in the terminology of [P.M.], "at least one propositional function formally equivalent to the propositional function $f\hat{z}$"). Russell does not explain his reasons for the form of the definition chosen, except for saying, correctly, that the definiens ought to be extensional; this, however, is likewise the case if a universal quantifier is used, as in 33-1.

The form of the definition with the existential quantifier seems to me not only to be less natural but also to lead to serious difficulties, which make

[2] [M.L.], § 26.

[3] Russell, "Mathematical Logic as Based on the Theory of Types", *American Journal of Mathematics*, XXX (1908), 222–62; for the definition, see p. 249.

[4] [P.M.], I, 71 ff., 187 ff.

[5] The definition in the original notation (p. 249 of the article mentioned above, and [P.M.], I, 76, 188) is as follows:

*20.01. $f\{\hat{z}(\psi z)\} \,.\!=:\, (\exists\phi) : \phi!x \,.\equiv_x .\, \psi x : f\{\phi!\hat{z}\}$ Df

Our transcription 33-2 is changed in inessential respects only. The exclamation point is omitted because it is necessary only on the basis of the ramified system of types, which is now generally regarded as unnecessary, and because it is at any rate inessential for the problem under discussion. The context is indicated only by dots instead of by a second-level variable, in order to make the definition applicable also to systems not containing such variables. The biconditional sign is used according to our abbreviation 3-1.

it appear doubtful whether the definition fulfils the purpose intended. To show this, let us consider two nonextensional properties of properties, say, Φ_1 and Φ_2, such that Φ_2 is the contradictory of Φ_1; hence Φ_2 holds in all cases, and only in those, in which Φ_1 does not hold. Since Φ_1 is nonextensional, there are different, but equivalent, properties, say f_1 and f_2, such that Φ_1 holds for f_1 and not for f_2, and hence Φ_2 holds for f_2. Then, according to definition 33-2, both Φ_1 and Φ_2 hold for the class $\hat{z}(f_1 z)$, although Φ_1 and Φ_2 are contradictories and hence logically incompatible. This would be an awkward result, although it does not constitute a formal contradiction, since Φ_1 and Φ_2 are logically exclusive only with respect to properties, while their application to classes is introduced merely as a certain mode of speech, which in the formal system itself, as distinguished from the informal interpretation in terms of classes, is merely a device of abbreviation.

In order to see the situation more clearly, let us try to construct a concrete example. As earlier (§ 26), let PM be the system constructed in [P.M.], and PM′ be the same system with some nonlogical constants added on the basis of rule 25-1. In order to find something like Φ_1 and Φ_2 in PM or PM′, we have to look for nonextensional signs. Among the very few such signs occurring in the system PM itself, there are the signs of identity '$=$' and nonidentity '$\neq$' when standing either between property expressions or between a property expression and a class expression. The sign '$=$' is actually used in [P.M.] in this way;[6] and the authors are aware that it is nonextensional in these contexts.[7] We shall first use the system PM′. We take as premises the following two sentences of this system:[8]

(i) '$(x)(\mathrm{F}x \bullet \mathrm{B}x \equiv \mathrm{H}x)$', or briefly, '$\mathrm{F} \bullet \mathrm{B} \equiv \mathrm{H}$'.
(ii) '$\mathrm{F}\hat{z} \bullet \mathrm{B}\hat{z} \neq \mathrm{H}\hat{z}$'.

These sentences say that the property Featherless Biped and the property Human are equivalent but not identical. Hence they are true. Now we shall examine the following two sentences:

(iii) '$\hat{z}(\mathrm{H}z) = \mathrm{H}\hat{z}$'.
(iv) '$\hat{z}(\mathrm{H}z) \neq \mathrm{H}\hat{z}$'.

[6] See [P.M.], I, 191, the proofs of *20.13 and *20.14.

[7] *Ibid.*, p. 84.

[8] For the convenience of the reader, we transcribe the notation of [P.M.] into our notation by writing the quantifier in the form '(x)' instead of in the form of a subscript and by using parentheses instead of dots. We keep, however, the notation '$\mathrm{H}\hat{z}$' for a property expression because this is an essential feature of the notation in [P.M.] (see above, § 26).

We shall expand these sentences by applying Russell's definition 33-2 in order to eliminate the class expression '$\hat{z}(Hz)$'. We substitute in this definition 'H' for 'f'; as '. . $\hat{z}(Hz)$ . .' we take (iii) and (iv) in turn. Thus (iii) is expanded into

(v) '$(\exists g)[(g \equiv H) \bullet (g\hat{z} = H\hat{z})]$'.

This sentence is provable in PM′, because it follows by existential generalization from the instance with 'H' for 'g'. Therefore, (iii) is provable and hence true on the basis of the interpretation assumed (§ 26). Now let us expand (iv). Here we have to take into consideration Russell's rule of context, according to which the smallest sentence or matrix in the actually given abbreviated notation is to be taken as corresponding to the left side in the definition 33-2. In other words, '$\neq$' is not to be eliminated before the elimination of the class expression, and hence the whole of (iv) is to be taken as '. . $\hat{z}(Hz)$ . .'.[9] Thus we obtain as the expansion of (iv):

(vi) '$(\exists g)[(g \equiv H) \bullet (g\hat{z} \neq H\hat{z})]$'.

This sentence is derivable from the conjunction of our premises (i) and (ii) by existential generalization with respect to '$F\hat{z} \bullet B\hat{z}$'. Hence, (iv) is derivable from the premises and therefore likewise true. Thus the result is that the sentences (iii) and (iv) are both true, although they look like contradictories. They do not actually constitute a contradiction because (iv) is not meant as the negation of (iii); this is shown by the fact that, according to the rules of the system PM′, (iv) is expanded not into '$\sim(\hat{z}(Hz) = H\hat{z})$' but into (vi). Nevertheless, our result shows that the notation of the system PM′ is here misleading, because it suggests the interpretation of (iv) as "$\hat{z}(Hz)$ is not identical with $H\hat{z}$", which would be in contradiction to (iii). It is true that Russell warns repeatedly that the class expressions are incomplete and have no meaning in isolation. On the other hand, the notation has been constructed with this aim in mind: The class expressions should be such that they can be manipulated as if they were names of entities; and Russell seems to assume that this aim has been reached.[10] Our result makes this assumption doubtful.

In the system PM itself, without the use of nonlogical constants, we can reach a similar result. Here we take as premise the assumption that there are two properties which are equivalent but nonidentical. Any par-

[9] It is stated ([P.M.], p. 188) that with regard to the scope of class expressions the same conventions are adopted as for descriptions. That the sign '$\neq$' when occurring in combination with a description is not eliminated before the elimination of the description is seen from the example in [P.M.], p. 173, line 2 from bottom.

[10] [P.M.], p. 188, ll. 3–5 and 14–16; and the text of p. 198.

ticular instance—for example, the conjunction of (i) and (ii)—can be formulated only in PM′, not in PM. But the existential assumption can be formulated in PM itself as follows:

(vii) '$(\exists g)(\exists f)[(g \equiv f) \bullet (g\hat{z} \neq f\hat{z})]$'.

In a way similar to the above we can derive from this premise in PM the following:

(viii) '$(\exists f)[(\hat{z}(fz) = f\hat{z}) \bullet (\hat{z}(fz) \neq f\hat{z})]$'.

This sentence is not provable in PM, but it is derivable from the premise (vii), which is, no doubt, true on the basis of the interpretation intended in [P.M.]; this work itself mentions the example of the properties Featherless Biped and Human. Although (viii) is not actually self-contradictory, still it looks as if it were. This shows again that the way the class expressions are introduced by Russell's definition is not quite in agreement with the intended purpose.

If, instead of Russell's definition 33-2, a definition involving a universal quantifier like 33-1 is used, then (iii) is not provable. In this case, both (iii) and (iv) are false. This apparently, but not actually, violates the principle of excluded middle; however, this seems less disturbing than the previous apparent violation of the principle of contradiction. If, furthermore, the rule of the context of the class expression is changed from Russell's form (the smallest sentence in the actually given abbreviated notation) to Quine's form (the smallest sentence in the primitive notation), as was done in 33-1, then (iv) is expanded into the negation of (iii). In this case, (iii) is false and (iv) is true, and thus there are no longer any puzzles. If in [P.M.] the definition of classes were changed according to 33-1, then only some of the proofs in a few subsections referring to the definition would need to be changed. It seems that later only extensional contexts occur; therefore, the theorems and proofs throughout the bulk of the work would remain unchanged.

4. Suppose that the language system *S* in question is such that every smallest matrix, that is, one which does not contain another matrix as a proper part, is extensional. This is the case, for instance, if modal operators are the only nonextensional signs. [Therefore, it is the case in S_2, where 'N' is the only nonextensional sign. Here, every nonextensional matrix contains a (proper or improper) part of the form 'N(. . .)' and hence a matrix '. . .' as a proper part. On the other hand, it is not the case in the system PM′ if we take it as including the sign '=' between property expressions. For example, the sentence '$H\hat{x} = H\hat{x}$' is of smallest size, but it is intensional (see above).] Then every class expression in *S* stands, after

the elimination of all other defined signs, within a smallest matrix which is extensional. Therefore, the class expression can here simply be replaced by the corresponding property expression, even if the smallest matrix in question stands within a wider nonextensional context. [For example, 'N[a ϵ $\hat{x}$(. . x . .)]' or 'N[$\hat{x}$(. . x . .)(a)]' is L-equivalent to, and hence L-interchangeable with, 'N[(λx)(. . x . .)(a)]', and hence also with 'N(. . a . .)'.] The reason for this is as follows. Let the smallest matrix containing a certain occurrence of '$\hat{x}$(Hx)' be represented by 'Φ($\hat{x}$(Hx))'. This is, according to our definition 33-1, L-equivalent to '$(g)[(g \equiv \text{H}) \supset \Phi g]$'. This obviously L-implies 'ΦH'; but the latter also L-implies the former (12-1), since, according to our assumption with respect to *S*, 'Φ' is extensional. Therefore, the two sentences are L-equivalent, and hence also L-interchangeable even in intensional contexts (12-2).

This shows that, in a system *S* of the kind described, we may simply take the property expressions themselves as class expressions also. This procedure is still simpler than procedure (2) explained above, which uses the contextual definition 33-1 for class expressions.

We have discussed four methods for the definition of classes in terms of properties. They can be used more generally for the definition of extensions of any kind in terms of intensions. These methods as here explained apply to symbolic object languages. The same methods can, of course, be applied in an analogous way to a word language and, in particular, to our metalanguage M. This latter application would be more important for us, because in our symbolic object languages we do not want to have class expressions in addition to property expressions, for the reasons explained earlier (§ 26), while in M we have phrases of both forms 'the class Human' and 'the property Human' and we should like to dispense with one of these forms in the primitive formulations in M. Since M contains identity sentences for properties (like PM'), it does not fulfil the condition required for *S* in method (4). But we could apply method (2) to M. This would consist in laying down the following three definitions; the first corresponds to 33-1, the second and the third are analogous to it:

33-3. . . . the class *f* . . . $=_{\text{Df}}$ for every property *g* equivalent to the property *f*, . . . the property *g* . . . .

33-4. . . . the truth-value *p* . . . $=_{\text{Df}}$ for every proposition *q* equivalent to the proposition *p*, . . . the proposition *q* . . . .

33-5. . . . the individual *x* . . . $=_{\text{Df}}$ for every individual concept *y* equivalent to the individual concept *x*, . . . the individual concept *y* . . . .

A convention determining the context indicated by dots would here be laid down similar to that for 33-1. (We may disregard here inessential changes of this context required by the accidents of idiom; for example, '*x belongs to* the class *f*' is changed to '*x has* the property *f*'.)

The three definitions here mentioned will not actually be adopted for M, because we shall find another, simpler form of a metalanguage which avoids even the apparent duplication of entities in M by entirely avoiding the duplication of expressions. This will be explained in the next section.

Would it be better to take properties as primitive and to define classes in terms of properties or to take classes as primitive and to define properties in terms of classes? We have explained four methods for the first alternative. Quine[11] rejects it for the reason that a property is even more obscure than a class. Which of the two is more obscure and which intuitively clearer is a controversial question. I shall not discuss this question here; it seems to be more psychological than logical. However, I think that most logicians agree that, if the terms 'class' and 'property' are understood in their customary sense, classes can be defined by properties, but it is hardly possible to define properties by classes (unless these classes are, in turn, characterized by properties); for a property determines its class uniquely, while many properties may correspond to a given class. It is, however, possible to define in terms of classes certain entities which stand in a one-one correlation to properties or other intensions and therefore may represent them for many purposes. We defined earlier the L-equivalence class of a designator in *S* as the class of all designators in *S* L-equivalent to it (3-15b). It is easily seen that there is a one-one correlation between the L-equivalence classes in *S* and the intensions expressible in *S*. Therefore, the L-equivalence class of a designator in *S* may be taken as its intension or at least as a representative for its intension. Procedures of this kind have been indicated by Russell and by Quine. Russell[12] mentions as a possibility the definition of a proposition as "the class of all sentences having the same significance as a given sentence". Quine[13] defines the meaning of an expression as the class of those expressions which are synonymous with it. Russell's concept of having the same significance and Quine's concept of synonymity correspond at least approximately to our concept of L-equivalence; if a stronger relation than L-equivalence is meant, for example, something like intensional isomorphism (§ 14), the concepts are, of course, analogous.

[11] [Notes], p. 126. [12] [Inquiry], p. 209. [13] [Notes], p. 120.

§ 34. The Neutral Metalanguage M′

While some symbolic systems (e.g., Russell's) have different expressions for properties and for classes, our systems (S_1 and S_2) have only one kind of expression. Analogously, we now introduce a "neutral" metalanguage M′. While M contains phrases like 'the property Human' and 'the class Human', M′ contains only the neutral expression 'Human'; and similarly with other types of designators. In this way the duplication of expressions in M is eliminated in M′, and thus the apparent duplication of entities disappears.

If, of the two phrases 'the class Human' and 'the property Human' in M, either the first were defined by the second or vice versa, then in the primitive notation of M we should have only one phrase instead of two, and hence the number of entities would be cut in half. I think that the same aim can also be reached in another and even simpler way. We have seen earlier (§ 26) that, on the basis of the method of extension and intension, the notation in a symbolic object language can be simplified. Instead of one expression as a name of a property (e.g., '$H\hat{x}$' in PM′) and another expression as a name of the corresponding class (e.g., '$\hat{x}(Hx)$'), it is sufficient to use one expression (e.g., '$(\lambda x)(Hx)$' or 'H' in S_1). This expression is, so to speak, neutral in the sense that it is regarded neither as a name of the property nor as a name of the class but rather as an expression whose intension is the property and whose extension is the class. If we apply an analogous procedure to the word language M, then our aim will be attained. Thus we have to look for a language form M′ in which we use, instead of the two phrases 'the class Human' and 'the property Human', only one phrase; this phrase, however, is not to be one of the two but rather another one which is neutral in containing neither the word 'class' nor the word 'property'. The simplest procedure is to take the word 'human' or 'Human' alone (the capitalized form to be used, as previously, at places where English grammar does not permit an ordinary adjective). We take M′ as the *neutral metalanguage* which results from M by these changes, that is, by eliminating the terms 'class', 'property', etc., in favor of neutral formulations. Our tasks is now to find suitable forms for formulations in M′. In this discussion we shall speak about M and M′, and hence we shall speak in a metametalanguage MM. For easier understanding, we take MM similar to M rather than to the less familiar M′; that is to say, we shall use terms like 'class', 'property', 'extension', 'intension', etc., in speaking about M′, although these terms cannot occur in M′ itself. The very next sentence will, in fact, be an example of this use. The term 'Human' in M′ is neutral in the same sense in which 'H' is neutral in S_1: 'Human' is regarded neither as a name of a class nor as a name of a property; it is, so to speak, at once a class expression and a property expression in the following way:

34-1. The extension of 'Human' in M′ is the class Human.
34-2. The intension of 'Human' in M′ is the property Human.

Analogously, instead of the two phrases 'the individual Scott' and 'the individual concept Scott' in M, we have in M′ the one neutral term 'Scott'. Here we have:

34-3. The extension of 'Scott' in M′ is the individual Scott.
34-4. The intension of 'Scott' in M′ is the individual concept Scott.

Since classes and properties have different identity conditions, a difficulty arises in the translation of identity sentences into M′. Take as an example the following sentences in M (see § 4):

34-5. The class Human is the same as the class Featherless Biped.
34-6. The property Human is not the same as the property Featherless Biped.
34-7. The property Human is the same as the property Rational Animal.

We translated, above, two phrases in M into 'Human' by simply omitting the words 'the class' and 'the property'. However, if we were to do the same with 34-5 and 34-6, a contradiction would obviously result. Generally speaking, since identity is different for extensions and intensions, a neutral formulation cannot speak about identity. Hence, identity phrases like 'is identical with' or 'is the same as' are not admissible in M′. How, then, to translate identity sentences into M′? Here the terms 'equivalent' and 'L-equivalent' in their nonsemantical use, as defined by 5-3 and 5-4, will help; note that in this use the terms stand for relations, not between designators, but between intensions. The definitions show that identity of extensions coincides with equivalence of intensions, and identity of intensions coincides with L-equivalence of intensions. Here in M′, the terms 'equivalent' and 'L-equivalent' can be used in connection with neutral phrases instead of phrases for intensions without any difficulty; therefore, we shall speak of equivalence and L-equivalence of neutral entities. Thus the general rules for the translation of identity sentences (in M or in a non-neutral object language, e.g., PM′) into neutral formulations in M′ are as follows:

34-8. A sentence stating identity of extensions is translated into M′ as a sentence stating equivalence of neutral entities.
34-9. A sentence stating identity of intensions is translated into M′ as a sentence stating L-equivalence of neutral entities.

Accordingly, we translate the identity sentences 34-5, 34-6, and 34-7 in M into the following sentences in M′:

34-10. Human is equivalent to Featherless Biped.
34-11. Human is not L-equivalent to Featherless Biped.
34-12. Human is L-equivalent to Rational Animal.

These three sentences can be obtained from 5-5, 5-6, and 5-7 in M by simply dropping the phrase 'the property'.

The sentences 34-10, 34-11, and 34-12 must be clearly distinguished from the following sentences, which look similar but are fundamentally different in their nature:

'Human' is equivalent to 'Featherless Biped' in M′.
'Human' is not L-equivalent to 'Featherless Biped' in M′.
'Human' is L-equivalent to 'Rational Animal' in M′.

These sentences are semantical sentences in MM concerning certain predicators in M′. Therefore, the predicators are included in quotation marks, and the sentences contain references to the language M′. They are perfect analogues to the sentences 3-8 and 3-11, which are semantical sentences in M (or M′) concerning predicators in S_1. On the other hand, the sentences 34-10, 34-11, and 34-12 are not semantical sentences; they do not speak about the predicators but use the predicators in order to speak about nonlinguistic entities. Therefore, the predicators are not included here in quotation marks, and there is no reference to a language system. The sentences belong to the nonsemantical (and, moreover, to the nonsemiotical) part of M′, to that part into which the sentences of the object languages can be translated. Sentence 34-10 is not only a translation of the sentence 34-5 in M, but also of the corresponding identity sentence 26-9 in PM′ and in ML′ (§ 25); 34-10 is, furthermore, an exact translation of the likewise neutral sentence '$H \equiv F \bullet B$' of S_1. Since 34-11 and 34-12 are intensional (in the sense of 11-3b), there cannot be sentences exactly corresponding to them in the extensional language S_1 (§ 11, Example IV). But there are such sentences in the modal language S_2, as we shall see later; thus '$H \equiv RA$' corresponds to 34-12.

Now let us see how neutral formulations of sentences are to be framed in M′. The translation of simple sentences, especially atomic sentences, into M′ involves no difficulty, since it corresponds closely to the customary formulation. Thus, for instance, as a translation of 'Hs' we take in M′ the simplest of the translations in M, namely, 'Scott is human' (which is 4-1). The other translations into M earlier discussed are not neutral (for instance, 4-2, 4-3, and those with 'individual' and 'individual concept' analogous to those mentioned in § 9); hence they are excluded from M′. The neutral formulation, 'Scott is human', in M′ replaces not only the non-neutral sentences in M just mentioned but also the two non-neutral phrases, 'the truth-value that Scott is human' and 'the proposition that Scott is human' in M (see 6-3 and 6-4). In some cases the simple formulation 'Scott is human' does not comply with ordinary English grammar, for instance, when occurring as a grammatical subject. In these cases we might, in analogy to 'Human', capitalize all words: 'Scott-Is-Human'; but this would be rather awkward for longer sentences. Another alternative is the addition of 'that' (see remark on 6-3 and 6-4): 'that Scott is human'. This formulation is to be used only as part of larger sentences, especially in the translation of sentences of M containing one of the phrases 'the truth-value that Scott is human' or 'the proposition that Scott is human'. In some cases, this formulation agrees with ordinary usage, in others not; but we shall admit it into M′ in all cases. Thus the (false) sentence 'N(Hs)' (in S_2) is the translated into 'It is necessary that Scott is human'. Since 'Hs' is equivalent to '(F • B)s', the following is true in M′:

34-13. That Scott is human is equivalent to that Scott is a featherless biped.

This formulation is admittedly somewhat awkward. The more customary formulations in M with 'the proposition' or 'the truth-value' inserted after 'to' are not possible here in M′ because they are not neutral; and there is no customary neutral noun. Therefore, we decide to admit the form 34-13 in M′, and likewise the analogous form 34-14 below.

Since 'Hs' is L-equivalent to 'RAs', the following is true in M′:

34-14. That Scott is human is L-equivalent to that Scott is a rational animal.

The use of 'equivalent' and 'L-equivalent' as nonsemantical terms standing between sentences, as in 34-13 and 34-14, is in analogy to the use of these terms between predicators (as in 34-10, etc.) and individual expressions, but here, between sentences, it is still more at odds with ordinary grammar. Fortunately, there is another formulation which is customary and grammatically correct; but it has the disadvantage that it is applicable only in connection with sentences, not with other designators. Instead of 'equivalent' we may use here 'if and only if', and instead of 'L-equivalent' 'that . . . if and only if - - -, is necessary'. (Here the phrase 'is necessary' is placed at the end only for the reason that English provides no other simple means to indicate that the argument of this phrase is the whole 'if and only if' sentence and not only its first component.) In this way, the following sentences take the place of 34-13 and 34-14:

34-15. Scott is human if and only if Scott is a featherless biped.

34-16. That Scott is human if and only if Scott is a rational animal, is necessary.

§ 35. M′ Is Not Poorer than M

The question is raised as to whether the designators in M′ are correctly described as neutral or whether they are, perhaps, actually names of intensions in disguise. If somebody wishes to regard 'Human' in M′ (or 'H' in S_2) as the name of a property, there is no essential objection. But it would be wrong to say that a language like S_2 or M′ contains only names of properties and no names of classes and therefore lacks important means of expression. Actually, all sentences of M are translatable into M′. That M′ is not poorer in means of expression than M is also shown by the possibility of reintroducing the non-neutral formulations of M into M′ with the help of contextual definitions.

Perhaps a reader who is accustomed to the usual method of the name-relation will have some doubts as to whether the language M′ or any other language can possibly be genuinely neutral; he will say that the allegedly neutral word 'Human' in M′, and likewise the corresponding sign 'H' in S_1 and S_2, in order to be unambiguous, must mean either as much as 'the property Human' or as 'the class Human'; in other words, it must be a name either of the property or of the class (compare Quine's comments below, § 44). I cannot quite agree with this either-or formulation. I think we should rather say that the word 'Human', and likewise 'H', stand both for the property as its intension and for the class as its extension. However, it must be admitted that the neutrality is not quite symmetrical. As we have seen earlier (§ 27), a designator stands primarily for its intension; the intension is what is actually conveyed by the designator from the speaker to the listener, it is what the listener understands. The reference to the extension, on the other hand, is secondary; the extension concerns the location of application of the designator, so that, in general, it cannot be determined by the listener merely on the basis of his understanding of the designator, but only with the help of factual knowledge. Therefore, if somebody insists on regarding a designator as a name either of its intension or of its extension, then the first would be more adequate, especially with respect to intensional languages like M′ and S_2. I think there is no essential objection against an application of the name-relation to the extent just described, for example, against regarding 'Human' in M′ and 'H' in S_2 as names of the property Human. The only reason I would prefer not to use the name-relation even here is the danger that this use might mislead us to the next step, which is no longer unobjectionable. In accordance with the customary conception of the name-relation, we might be tempted to say: "If 'Human' (or 'H') is a name for the property Human, where do we find a name for the class Human? We wish to speak, not only about properties, but also about classes; therefore, we are not satisfied with a language like M′ or S_2, which does not provide names for classes and other extensions." This I should regard as a misconception of the situation. *M′ is not poorer than M* by not containing the phrase 'the class Human'. Whatever is expressed in M with the help of this phrase is translatable into M′ with the help of 'Human'; and whatever is expressed in a non-neutral symbolic language like PM′ with the help of the class expression '$\hat{x}(\mathrm{H}x)$' is translatable into S_2 with the help of 'H'. The simplest method for the translation into S_2 is based on the method (4) explained in the preceding section. For the sake of an example, let us take, not the system PM′, but the system PM″, which is like PM′ except for containing our form of the contextual definition of classes (33-1) instead of Russell's (33-2). The rules of translation from PM″ into S_2 with respect to class expressions are as follows:

35-1. a. For the translation of a smallest sentence (or matrix) which is extensional and does not contain '=', both a property expression (e.g., '$\mathrm{H}\hat{x}$) and a class expression (e.g., '$\hat{x}(\mathrm{H}x)$') in PM″ are translated into S_2 by the corresponding neutral expression (e.g., 'H' or '$(\lambda x)(\mathrm{H}x)$'). (This rule is based on method (4), explained in § 33.)

b. An identity sentence in PM″ with two class expressions (e.g., '$\hat{x}(\mathrm{H}x) = \hat{x}(\mathrm{F}x \bullet \mathrm{B}x)$') is translated into an ≡-sentence with the corresponding neutral expressions (e.g., '$\mathrm{H} \equiv \mathrm{F} \bullet \mathrm{B}$').

c. An identity sentence with two property expressions (e.g., '$\mathrm{H}\hat{x} = \mathrm{RA}\hat{x}$') is translated into the corresponding ≡-sentence (e.g., '$\mathrm{H} \equiv \mathrm{RA}$').

(We leave aside here identity sentences with one class expression and one property expression; all such sentences are L-false.)

The translation from M into M′ is analogous. We may assume that any sentence of smallest size in M which is not an identity sentence is extensional. Then the rules are as follows:

35-2. a. In all contexts except identity sentences, both class expressions (e.g., 'the class Human') and property expressions (e.g., 'the property Human') are translated by the corresponding neutral expressions (e.g., 'Human').

b. A sentence stating the identity of classes is translated into a sentence stating the equivalence of the corresponding neutral entities.

c. A sentence stating the identity of properties is translated into a sentence stating the L-equivalence of the corresponding neutral entities.

Rules 35-2b and c are special cases of the general rules 34-8 and 34-9 for extensions and intensions (see examples 34-10, 34-11, and 34-12).

Thus we see that the view that M′ is poorer than M is a misconception. Since the formulation "The designators in M′ are names for intensions, and there are no names for extensions in M′", may easily lead to this misconception, it seems to me inadvisable. It seems more adequate and less misleading to say either "every designator in M′ has an intension and an extension" or "the designators in M′ are neutral".

In the translations by rule 35-2a, characterizing phrases like 'the class', 'the property', etc., are simply dropped. This might give the impression, perhaps, that in the transition from M to M′ certain important distinctions disappear. This, however, is not the case. All the distinctions made in M are preserved in M′; they are only formulated in a different and, in general, in a simpler way. This is shown by the fact that all the non-neutral ways of speaking in M with terms like 'class', 'property', etc., could be reintroduced into M′ by contextual definitions if we wanted them there. (In fact, of course, we do not want to destroy the neutrality of M′.) Thus the terms 'class' and 'property' could be introduced by the following contextual definitions:

35-3. a. . . . the class *f* . . . $=_{\mathrm{Df}}$ for every *g*, if *g* is equivalent to *f*, then . . . *g* . . . .

b. . . . the property *f* . . . $=_{\mathrm{Df}}$ for every *g*, if *g* is L-equivalent to *f*, then . . . *g* . . . .

(Concerning the context indicated by dots, see the remarks following 33-5.) If the context indicated by dots is extensional, we may take, instead of (a), the simpler definition:

a′. . . . the class *f* . . . $=_{\mathrm{Df}}$ . . . *f* . . . .

If the context is either extensional or intensional, we may take, instead of (b), the simpler form:

b′. . . . the property *f* . . . $=_{\mathrm{Df}}$ . . . *f* . . . .

Contextual definitions for the terms 'individual', 'individual concept', 'truth-value', and 'proposition' are analogous. It is admitted that these definitions lead in some cases to unusual formulations. However, they do not lead to false results. The decisive point is that they also yield the original non-neutral formulations in M.

Identity sentences like those in M can likewise be reintroduced into M′ by a procedure the reverse of that described in 34-8 and 34-9:

35-4. **a.** The class *f* is the same as the class *g* $=_{\mathrm{Df}}$ *f* is equivalent to *g*.

b. The property *f* is the same as the property *g* $=_{\mathrm{Df}}$ *f* is L-equivalent to *g*.

The possibility of these definitions in M′ for the non-neutral formulations in M shows that all distinctions in M are actually preserved in M′ in a different form. In other words, M′ is not poorer in means of expression than M.

§ 36. Neutral Variables in M′

Some symbolic systems have different variables for classes and for properties; we have seen earlier (§ 27) that this is unnecessary. Similarly, the phrases 'for every class' and 'for every property' in M constitute an unnecessary duplication. They are replaced in M′ by 'for every *f*', where '*f*' is a neutral variable whose value-intensions are properties and whose value-extensions are classes. Neutral variables for other types are introduced analogously.

There are still other non-neutral expressions in M which have to be replaced by neutral expressions in M′, namely, those phrases by which we refer in a general way to entities of some kind, for instance, pronouns like 'every', 'any', 'all', 'some', 'there is', 'none', in combination with words like 'class', 'property', etc. In a symbolic language, phrases of this kind are translated with the help of variables in quantifiers. We have seen earlier (§ 27) that in a symbolic language not only the use of different expressions for classes and for properties is an unnecessary duplication, but so is likewise the use of different variables for classes and for properties (as, for instance, 'α' and 'ϕ' in the system PM). Instead, we may use *neutral variables*, whose value-extensions are classes and whose value-intensions are properties. Now we shall do the same in M′, in order to make possible the neutral formulation of general sentences. We supplement the word language in M′ by the following letters as variables: '*f*', '*g*', etc., for predicators of level one and degree one as value expressions; '*x*', '*y*', etc., for individual expressions; '*p*', '*q*', etc., for sentences. Thus, a non-neutral formulation of a universal sentence in M containing one of the two phrases 'every class' (or 'all classes') and 'every property' (or 'all properties') is translated into a neutral sentence of M′ with the help of the phrase 'for every *f*', corresponding to a universal quantifier. Likewise, an existential sentence in M containing one of the phrases 'some class' (or 'there is a class') and 'some property' (or 'there is a property') is translated into M′ with the help of 'for some *f*' (or 'there is an *f*'). Analogously, a general sentence concerning propositions or truth-values is translated into M′ with the help of 'for every *p*' or 'for some *p*' (or 'there is a *p*'). And a general sentence concerning individuals or individual concepts is translated with the help of 'for every *x*' or 'for some *x*' (or 'there is an *x*'). Examples will be given later (see 43-4).

Universal sentences in M about extensions or intensions in general can likewise be translated into M′ if we introduce general variables for which designators of all types are value expressions. To avoid contradictions, suitable restrictive rules have to be laid down for the use of these general variables; this can be done in different ways.[14]

§ 37. On the Formulation of Semantics in the Neutral Metalanguage M′

Two semantical relations between expressions and neutral entities, designation and L-designation, are introduced into M′. It is shown how semantical rules and statements in M can then be translated into M′. The relation of designation is extensional; it is used for the translation of statements concerning the extension of given expressions. The relation of L-designation is nonextensional; it serves for the translation of statements concerning the intension of expressions. Thus the whole semantics of a system (e.g., S_1) can be translated from M into M′.

In the preceding sections we have discussed only the nonsemantical part of the metalanguage, that part into which the sentences of object languages can be translated. We come now to the more important semantical part of the metalanguage, that part in which we speak about the sentences and other expressions of the object languages, applying to them semantical terms like 'true', 'L-true', 'equivalent', 'L-equivalent', etc. Most of the discussions in the earlier chapters of this book are formulated in this semantical part of the metalanguage M. This holds, in particular, for those statements which speak about classes, properties, propositions, etc., not only in a general way but in relation to expressions of an object language—for instance, the following two (§ 4):

37-1. The extension of 'H' in S_1 is the class Human.

37-2. The intension of 'H' in S_1 is the property Human.

[14] For a historical survey of different methods of avoiding the antinomies, see Quine, [M.L.], § 29.

The important question now is whether it is possible also to translate these semantical statements of M into the neutral metalanguage M′, that is to say, into formulations which, instead of phrases like 'the class Human' and 'the property Human', use only neutral phrases like 'Human'. Only if this is possible can we say that we have overcome the duplication of entities.

We shall see that it is indeed possible to translate semantics from M into M′. The sentence 37-1 states that the relation of extension holds between the class Human and the predicator 'H' (in S_1), and 37-2 states that the relation of intension holds between the property Human and the same predicator. How can we obtain neutral formulations in M′ referring to the neutral entity Human instead of to the class and the property? It would, of course, not do simply to drop the phrases 'the class' and 'the property' in those sentences, because then the same entity would be asserted to be at once the extension and the intension of the same predicator, and that would not be in accordance with the intended meaning of the terms 'extension' and 'intension'. Instead, we must make use of a relation which holds between the neutral entity Human and the predicator 'H', a relation which can be neither the relation of extension nor that of intension, although it is similar to them. A closer investigation of the situation shows that we need here two new relations, both holding between 'H' and Human; the first of them is related to the second as a radical semantical concept (e.g., truth) to the corresponding L-concept (e.g., L-truth). Therefore, it seems natural, if we find a suitable word for the first relation, to take the same word with the prefix 'L-' for the second. The first relation is here meant in such a way that it is definable also in an extensional metalanguage; but the second relation is intensional, as we shall see. Since the first relation holds between an expression (e.g., 'H') and an entity (e.g., Human) for which that expression stands, a word like 'means', 'signifies', 'expresses', 'designates', 'denotes', or something similar would seem suitable. I do not wish to make a specific suggestion. Let us tentatively use the term 'designates' for the first relation, and hence 'L-designates' for the second. Then, instead of 37-1 and 37-2 in M, we have in M′, with respect to S_1, the following:

37-3. 'H' designates Human.

This may be regarded as the formulation in M′ of a rule of designation for the system S_1 (corresponding to the first item in the previous rule 1-2). The first relation is intended to be extensional; that is to say, any full sentence of it is extensional with respect to each of the two argument expressions. Hence, 37-3 is extensional with respect to 'Human'; that is to say, the occurrence of 'Human' in this sentence is interchangeable with any predicator which is equivalent to 'Human' in M′. Thus we obtain the following two results, according to the equivalences stated in § 34:

37-4. 'H' designates Featherless Biped.

37-5. 'H' designates Rational Animal.

By using a neutral predicator variable '*f*' (see § 36) and 'equivalent' as a nonsemantical term (see 5-3 and § 34), we can express the result in a general form:

37-6. For every *f*, if *f* is equivalent to Human, then 'H' designates *f* (in S_1).

If a suitable definition for 'designates in S_1' is laid down, which has not been done here, then the converse of 37-6 also holds:

37-7. For every *f*, 'H' designates *f* (in S_1) if and only if *f* is equivalent to Human.

We have decided to use the term 'L-designates' for the second relation. We shall not give a definition for it. We assume for the following discussion that it is defined with respect to a given system, say S_1, in such a way that the following condition 37-8 is fulfilled; an analogous condition holds for L-truth, according to our convention 2-1, and for the other L-concepts.

37-8. An expression $\mathfrak{A}_i$ *L-designates* an entity *u* in S_1 if and only if it can be shown that $\mathfrak{A}_i$ designates *u* in S_1 by merely using the semantical rules of S_1, without any reference to facts.

(The variable '*u*' here used in M′ is a general, that is, not type-restricted, variable; see the remarks at the end of § 36.) Now let us apply 37-8 to 37-3, 37-4, and 37-5 in turn. Statement 37-3 can be established on the basis of the semantical rules of S_1 alone in a trivial way, since it is itself one of these rules. This yields, with respect to S_1:

37-9. 'H' L-designates Human.

The same, however, does not hold for 37-4. In order to show that this statement holds, we have used and must use not only the semantical rule 37-3 but also the result that the predicators 'Human' and 'Featherless Biped' are equivalent in M′; this equivalence, like that of the corresponding predicators 'H' and 'F • B' in S_1, is not an L-equivalence (see § 34) but is based on biological fact (3-6). Hence, according to 37-8, the following is true in M′:

37-10. 'H' does not L-designate Featherless Biped.

Since 'Human' and 'Featherless Biped' are equivalent in M′, we see from 37-9 and 37-10 that the relation of L-designation is nonextensional.

Statement 37-5 can again be established on the basis of rule 37-3 alone, without reference to facts, because 'Human' and 'Rational Animal' are supposed to mean the same (see remark on 1-2). Hence, according to 37-8, the following is true:

37-11. 'H' L-designates Rational Animal.

We can formulate the result in a general form with a neutral variable '*f*' and 'L-equivalent' as a nonsemantical term:

37-12. For every *f*, if *f* is L-equivalent to Human, then 'H' L-designates *f*.

If a suitable definition for 'L-designates' is laid down in accordance with the convention 37-8, then also the converse of 37-12 holds:

37-13. For every *f*, 'H' L-designates *f* if and only if *f* is L-equivalent to Human.

Statement 37-3 may be regarded as a translation of 37-1 into M′, and likewise 37-9 as a translation of 37-2. It is true that the explicit reference to a class in 37-1 is not directly mirrored by any expression in 37-3, but it is indirectly represented by the extensionality of 37-3 with respect to 'Human', which is shown by the instance 37-4 and generally by 37-6. Thus, 37-6 may also be regarded as a translation of 37-1. Similarly, the explicit reference to a property in 37-2 is indirectly represented by the intensionality of 37-9 with respect to 'Human', which is exhibited in instances like 37-10 and 37-11 and generally in 37-12. Thus, 37-12 may also be regarded as a translation of 37-2.

We have shown the application of the relations of designation and L-designation to predicators. The application to designators of other types is quite analogous. As examples with respect to individual expressions in S_1, in analogy to 37-3, 37-4, 37-6, and 37-7, the following sentences are true in M′:

37-14. 's' designates Walter Scott.

37-15. 's' designates The Author Of Waverley.

37-16. For every *x*, if *x* is equivalent to Walter Scott, then 's' designates *x*.

37-17. For every *x*, 's' designates *x* if and only if *x* is equivalent to Walter Scott.

Rule 37-14 is a rule of designation of the system S_1, corresponding to the first item in 1-1. Sentence 37-15 is derived from 37-14 with the help of a historical fact (9-1). Further, in analogy to 37-9, 37-10, 37-12, and 37-13, the following sentences are true in M′:

37-18. ‘s’ L-designates Walter Scott.
37-19. ‘s’ does not L-designate The Author Of Waverley.
37-20. For every *x*, if *x* is L-equivalent to Walter Scott, then ‘s’ L-designates *x*.
37-21. For every *x*, ‘s’ L-designates *x* if and only if *x* is L-equivalent to Walter Scott.

Sentences 37-14 and 37-16 may be regarded as translations of the following sentence in M (§ 9):

‘The extension of ‘s’ is the individual Walter Scott’.

Sentences 37-18 and 37-20 may be regarded as translations of:

‘The intension of ‘s’ is the individual concept Water Scott’.

Remarks analogous to those made above on 37-3 and 37-9 hold here.

Analogously, with respect to sentences in S_1, the following statements are true in M′; we use a that-clause for the neutral formulation (§ 34):

37-22. ‘Hs’ designates that Scott is human.

This statement, in distinction to 37-3 and 37-14, is itself not a semantical rule but follows from these rules with the help of a suitable definition for ‘designates in S_1’, as applied to sentences. The following is a consequence of 37-22, because ‘Scott is human’ and ‘Scott is a featherless biped’ are equivalent in M′:

37-23. ‘Hs’ designates that Scott is a featherless biped.

Generally, with the neutral variable ‘*p*’ (§ 36):

37-24. For every *p*, if *p* is equivalent to that Scott is human, then ‘Hs’ designates *p*.

37-25. For every *p*, ‘Hs’ designates *p* if and only if *p* is equivalent to that Scott is human.

(In these two statements, the nonidiomatic phrase ‘is equivalent to that’ may be replaced by ‘if and only if’; see the explanations to 34-13 and 34-15.)

Furthermore, for L-designation, the following statements are true in M′:

37-26. ‘Hs’ L-designates that Scott is human.
37-27. ‘Hs’ does not L-designate that Scott is a featherless biped.
37-28. For every *p*, if *p* is L-equivalent to that Scott is human, then ‘Hs’ L-designates *p*.

37-29. For every p, 'Hs' L-designates p if and only if p is L-equivalent to that Scott is human.

(In the last two statements, the nonidiomatic phrase 'is L-equivalent to that' can be avoided by a transformation analogous to that of 34-14 into 34-16.)

Sentences 37-22 and 37-24 may be regarded as translations of the sentence 6-3 in M concerning the truth-value as extension; likewise, 37-26 and 37-28 as translations of the sentence 6-4 concerning the proposition as intension. Remarks analogous to the earlier ones hold here.

We have previously seen that it would be possible to reintroduce the non-neutral terms 'class', 'property', etc. into M′ by contextual definitions. If we were to apply these terms in the formulation of semantical statements in M′, these statements would become quite similar to those in M. For example, by applying the definition of 'class' (35-3a) to 37-6, we obtain:

37-30. 'H' designates the class Human.

Likewise, by applying the definition of 'property' (35-3b) to 37-12, we obtain:

37-31. 'H' L-designates the property Human.

Analogous results would be obtained for individual expressions and sentences. These results show that the relation of designation in M′ corresponds to the relation between a designator and its extension in M, and the relation of L-designation in M′ corresponds to the relation between a designator and its intension in M.[15]

[15] My use of the terms 'designation' and 'designatum' in [I] was, as I realize now, not quite uniform, because at that time I did not yet see clearly the distinction which I make now in M with the help of the terms 'extension' and 'intension', and in M′ with the help of the terms 'designation' and 'L-designation'. The use of 'designatum' in [I] corresponds in most cases to the present use of 'intension' in M (or 'L-designatum' in M′). Thus, in the Table of Designata ([I], p. 18) and in later examples of Rules of Designation, the following kinds of entities are taken as designata: properties, relations, attributes, functions, concepts, and propositions. It is only with respect to individual expressions that I used the term in a different way, taking as designata in the table and in the examples not individual concepts but individuals. Since it is not customary to speak of individual concepts under any term, I was not aware of the fact that they, and not individuals, belong to the same category as properties, propositions, etc. Thus, in the case of individual expressions, what I took as designata were the same as what would be taken as nominata by the method of the name-relation. It is probably due to this fact that Church ([Review C.]) understood my term 'designatum' in all cases in the sense of 'nominatum'; and presumably Quine ([Notes]) likewise believes himself to be in accord with my use when he applies 'designatum' in this sense. I regret that the lack of a clear explanation in [I] has caused these misunderstandings. This lack was not accidental but was caused by an obscurity of long standing in some of the fundamental semantical concepts. If I see it correctly, this obscurity has been overcome only by the analysis made in this book. Church's statement ([Review C.], pp. 299 f.) that the designatum of a sentence is not a proposition but a truth-value is—on the basis of Frege's method of the name-relation—correct for Church's use of

The examples in this section show how semantical sentences in M, stating the extensions or intensions of predicators, individual expressions, and sentences in S_1, can be translated into neutral formulations in M′. The translation of semantical sentences which refer not to nonlinguistic entities but only to expressions in the object language, for instance, sentences about truth, L-truth, equivalence, and L-equivalence, does, of course, not involve any difficulty. Thus the whole of semantics, with respect to S_1 or any other system, can be translated from M into M′.

The reasons for our use of the two metalanguages, M and M′, may be briefly summarized. Metalanguage M was used in the first three chapters of this book in an uncritical way, so to speak. It supplies pairs of terms 'class'-'property', and the like, and the general terms 'extension' and 'intension'. The use of these terms constituted what we have called the method of extension and intension. The chief reason for using these pairs of terms is that they correspond to familiar concepts, usually regarded as kinds of entities. In the present chapter we constructed the neutral metalanguage M′, which has no such pairs of terms and thus avoids the appearance of a duplication of entities. Although the terms 'extension' and 'intension' do not occur in M′, the essential features of the method used in M′ are still the same as in M; therefore, we might still call the method used in M′ the neutral form of the method of extension and intension, or else the (neutral) method of equivalence and L-equivalence, or the (neutral) method of designation and L-designation. The distinctions made in M are not neglected in M′ but are represented in a different form. Instead of an apparent duplication of entities, we have here a distinction between two relations among expressions, namely, equivalence and L-equivalence, and, based upon it, a distinction between two relations between expressions and entities, namely, designation and L-designation. We have seen that it is possible to construct in M′ contextual definitions for the non-neutral terms 'class', 'property', etc., which lead to formulations like those in M. This result shows, on the one hand, that the neutral method in M′ does indeed preserve all distinctions originally made in M and hence is an effective substitute for the original form of the method. On the other hand, the result is a justification for M, since it shows that the

---

'designatum' in the sense of 'nominatum'; not, however, for my use of 'designatum' in [I] in the sense of 'intension'.

In [I], I occasionally used the terms 'synonymous' and 'L-synonymous'. The distinction which I had in mind but did not grasp satisfactorily is now expressed more adequately by the terms 'equivalent' and 'L-equivalent' in their application to designators in general.

apparent duplication of entities in M is, in fact, only a duplication of modes of speech.

Since the non-neutral mode of speech in M and the neutral mode of speech in M′ cover the same domain, the choice between them is a matter of practical preference. The neutral formulation is much simpler and avoids even the appearance of a duplication of entities. Therefore, this formulation might be preferable in cases in which a metalanguage for semantical purposes is to be constructed in a strict, systematized way, for instance, in a symbolic language or in words whose use is regulated by explicit rules. On the other hand, the non-neutral formulation is in most cases more familiar, more in accordance with ordinary usage. Therefore, this formulation may seem preferable for semantical discussions which are not on a highly technical level, especially for purposes of introductory explanations. That is the reason for its use in the first part of this book.

§ 38. On the Possibility of an Extensional Metalanguage for Semantics

The question is discussed as to whether a complete semantical description of a system, even a nonextensional system like S_2, can be formulated in an extensional metalanguage, for instance, the sublanguage M_e of M′ containing only the extensional sentences of M′. It is found that most of the semantical rules (rules of formation, of truth, and of ranges) can be formulated in M_e without any difficulty. The situation is not so simple with respect to the rules of designation; but it seems that these rules can also be adequately formulated in M_e.

We have formulated semantical sentences in two different metalanguages, M and M′. Both these languages are nonextensional. The question arises as to whether semantics can be formulated in an extensional metalanguage—more exactly, whether it is possible to construct an extensional metalanguage sufficient for the formulation of a complete semantical description even of a nonextensional object language (as, for instance, S_2). A semantical description of an object language is complete if it, given as the only information about the language, enables us to understand every sentence of the language and hence to determine whether or not it is L-equivalent to any given sentence of our metalanguage. The answer to the question is not at present known. However, on the basis of some studies I have made, an affirmative answer seems to me not improbable. Here I shall give a few indications only.

It is easily seen that a sentence in M which says what the intension of a certain expression is, is nonextensional. For example, the sentence 'the intension of 'H' in S_1 is the property Human' (4-17) is nonextensional with respect to 'the property Human', because if this predicator is replaced by the equivalent one, 'the property Featherless Biped', then the true sentence is changed into a false one. Sentences of this kind are essential for the use of our method in M. Therefore, if we wish to find extensional semantical sentences, it seems more promising to look at the neutral formulations in M′. The term 'intension' does not occur in M′; nor do those intensional sentences of M which state the identity or the nonidentity of properties or other intensions (for instance, 4-8 and 4-9). Nevertheless, M′ is not extensional; the semantical formulations which we used in M′ contain the following three nonextensional (and, moreover, intensional) terms and no others. The first is the modal term 'necessary' (see, for instance, 34-16). The second is the term 'L-equivalent' in its nonsemantical use, as occurring, for instance, in 34-11, 34-12, and 34-14; it is easily seen that each of these sentences is nonextensional with respect to both argument expressions. This term is definable on the basis of 'necessary' (compare, for instance, 34-14 and 34-16). [Note, incidentally, that the semantical term 'L-equivalent in the system *S*' is extensional. For example, '$\mathfrak{S}_1$ is L-equivalent to $\mathfrak{S}_2$ in the system S_1' is extensional; in contradistinction to 34-14, it does not contain sentences as parts, but only names of sentences.] The third nonextensional term in M′ is 'L-designates' (see the remark following 37-10).

Let M_e be the metalanguage which contains all the extensional sentences of M′ and no others; we can construct it out of M′ by omitting all sentences containing the three nonextensional terms mentioned. Our question is: How much of the semantics, say of the extensional system S_1 and the intensional system S_2, can be formulated in M_e?

A complete system of semantical rules for S_1 or S_2, which is not given in this book, would consist of the following kinds of rules:

(i) Rules of formation, on the basis of a classification of the signs; these rules constitute a definition of 'sentence'.
(ii) Rules of designation for the primitive descriptive constants, namely, individual constants and predicates.
(iii) Rules of truth.
(iv) Rules of ranges.

It is easy to see that the rules of kinds (i), (iii), and (iv) can be formulated in an extensional metalanguage like M_e. We must here consider these rules in their exact formulation. The designations of expressions of the object language must be formed, not with the help of quotation marks, as we did for the sake of convenience in the previous examples of semantical rules and statements, but as descriptions with the help of German letters.

Let us add here, for this purpose, the letter '$\mathfrak{N}$' as designation in M_e of the modal sign 'N' in S_2. As an example of a rule of formation for S_2 in M_e, let us take the rule for N-matrices: 'If $\mathfrak{A}_i$ is a matrix in S_2, then $\mathfrak{N}(\mathfrak{A}_i)$ is a matrix in S_2.' In application to the instance 'Hs', this rule says that, if 'Hs' is a matrix, as it is, indeed, according to another rule, then 'N(Hs)' is a matrix. Note, however, that the rule itself does not contain the expression $\mathfrak{A}_i$, for instance, 'Hs', but only refers to this expression by using a name '$\mathfrak{A}_i$' for it (actually, a variable for which a name, say '$\mathfrak{A}_1$', may be substituted). Among the rules of truth we leave aside for the moment that for atomic sentences because it contains the term 'designates' (or 'refers to', see 1-3) which will be discussed later. The following is an example of one of the other rules of truth (1-5): 'A disjunction of two sentences $\mathfrak{S}_i$ and $\mathfrak{S}_j$ (that is to say, a sentence consisting of $\mathfrak{S}_i$ included in parentheses followed by the wedge followed by $\mathfrak{S}_j$ included in parentheses) is true if and only if either $\mathfrak{S}_i$ or $\mathfrak{S}_j$ or both are true.' It is clear that this formulation is extensional. The same holds for the rules of ranges for S_2, which will be given in § 41. These rules define 'the sentence $\mathfrak{S}_i$ holds in the state-description $\mathfrak{R}_n$'; $\mathfrak{R}_n$ is a class of sentences. Note that the sentence $\mathfrak{S}_i$, let alone the class $\mathfrak{R}_n$, does not itself occur in the rule; only the names (or variables) '$\mathfrak{S}_i$' and '$\mathfrak{R}_n$' occur. Thus it is clear that the relation of holding is extensional. The rules of ranges refer, moreover, to assignments; an assignment is a function which assigns to a variable and a state-description as arguments an individual constant as value. Only the extensions of these functions are essential for the rules and the statements based upon the rules; that is to say, if a reference to one assignment in a true statement is replaced by a reference to another equivalent assignment (i.e., one which assigns to all pairs of arguments the same values as the first assignment), then the resulting statement is likewise true. Note, further, that the exact formulation of the rule concerning 'N' (41-2g) has the form: 'A matrix $\mathfrak{N}(\mathfrak{A}_i)$ holds . . .'; thus it does not contain the modal sign 'N' itself but only its name '$\mathfrak{N}$'. Thus we see that all rules of ranges for S_2, including the rule concerning 'N', are extensional.

Now we go back to the rules of designation. Here is the one critical point for our problem of the expressibility of the semantics of S_2 in M_e. In M′, we distinguished two relations between designators and neutral entities, namely, designation and L-designation. The relation of designation is extensional and hence does occur in M_e; but the relation of L-designation does not. Thus we have to examine the question as to whether the relation of designation suffices for describing the meanings of the expressions in the object languages. For instance, the meaning of 'H' in S_1

and S_2 is (the property) Human, not Featherless Biped; the meaning of 's' is Walter Scott, not The Author Of Waverley. In M′, we can easily express this distinction with the help of the term 'L-designation' by the statements 37-9 and 37-10, 37-18 and 37-19. But how can we do it in M_e, where we have only the term 'designation'? The difficulty consists in the fact that, with respect to designation, the following two statements are both true (37-3 and 37-4):

38-1. 'H' designates Human.
38-2. 'H' designates Featherless Biped.

And the same holds for the following two statements (37-14 and 37-15):

38-3. 's' designates Walter Scott.
38-4. 's' designates The Author Of Waverley.

In view of this fact, it might seem at first glance as though it were impossible to give in M_e the information about the meanings intended for 'H' and 's'. However, I believe that this is not impossible. In M_e we lay down 38-1 and 38-3 among the rules of designation for S_1 and S_2. Then the statement 38-2, although it is likewise true, is fundamentally different from 38-1, for it is neither a semantical rule, nor derivable from the semantical rules alone; it was derived from rule 38-1 together with a biological fact (3-6). If the metametalanguage MM, in which we are speaking here about M_e and the other metalanguages, contains L-terms, then we can formulate the difference in this way: 38-1 is L-true in M_e but 38-2 is only F-true. The relation between 38-3 and 38-4 is analogous. But even in M_e itself we can describe the situation in more explicit terms. If we wish to add to 38-1 a negative statement in M_e, the following may be taken (3-8):

38-5. 'H' and 'F • B' are not L-equivalent (in S_1 and S_2).

This statement, together with 38-1 and some other semantical rules, corresponds in a certain sense to the negative statement 37-10 in M′.

The rules of designation themselves refer only to the primitive individual constants and predicator constants. But the extensional relation of designation can also be defined in M_e in a wider sense so as to apply to all designators, including compound individual expressions, predicators, and sentences, also intensional sentences in S_2. Then, for example, the following two statements hold in M_e (37-22 and 37-23):

38-6. 'Hs' designates that Scott is human.
38-7. 'Hs' designates that Scott is a featherless biped.

The difference between these two statements is analogous to that between 38-1 and 38-2: Statement 38-6, though not itself a rule, follows from the semantical rules alone, while for the derivation of 38-7 a factual premise is needed.

The foregoing discussion shows that, even if somebody possesses no other information concerning S_1 and S_2 than the semantical rules for these systems formulated in M_e, he is, nevertheless, in a position to know the meanings—that is to say, not only the extensions but also the intensions—which are intended, first, for the primitive descriptive constants and, second, for all designators. All he has to do is to look, first, at the rules of designation themselves and, second, at those statements about designation which follow from the semantical rules alone, leaving aside all those statements in M_e which, although true, can be arrived at only with the help of factual knowledge. In other words, he has to consider only those statements about designation which are L-true in M_e.

It is sometimes said that a metalanguage, in which the semantics of an object language *S* is to be formulated, must contain translations of all expressions or at least of all designators in *S*. If this were right, M_e would not suffice as a semantics language for S_2, because M_e cannot, of course, contain an expression L-equivalent to the intensional sign 'N' in S_2. But the requirement mentioned is only approximately right; strictly speaking, it is too strong. The metalanguage must, indeed, contain for every sentence in *S* an L-equivalent sentence; furthermore, it must be sufficiently equipped with variables and descriptive expressions. It is, however, not necessary that it contain an L-equivalent expression for every logical sign in *S*. Although M_e cannot contain a translation of 'N', it can contain a semantical rule for 'N', for instance, the rule of ranges mentioned above. If $\mathfrak{S}_i$ is a sentence in S_2 containing 'N', then an extensional language like S_1 or M_e cannot, of course, contain a translation of $\mathfrak{S}_i$ in the strong sense of a sentence with the same intensional structure (§ 14). But it can be shown that S_1, and hence M_e, too, always contains a sentence L-equivalent to $\mathfrak{S}_i$. [For full sentences of 'N', this follows simply from the circumstance that they are either L-true or L-false (see 39-2); however, since sentences may contain several occurrences of 'N' and quantifiers in any combination, the general proof is rather complicated.] Further, S_1 and S_2 contain the same variables and descriptive signs. Hence, if M_e is sufficient for the formulation of the semantics of S_1, it is likewise sufficient for that of S_2.

On the basis of these considerations, I am inclined to believe that it is possible to give a complete semantical description even of an intensional language system like S_2 in an extensional metalanguage like M_e. However, this problem requires further investigation.

CHAPTER V

ON THE LOGIC OF MODALITIES

In this chapter we study logical modalities like necessity, possibility, impossibility. We introduce 'N' as a symbol of necessity; the other modal concepts, including necessary implication and necessary equivalence, can be defined with its help. The modal system S_2 is constructed by adding 'N' to our previous system S_1 (§ 39); and the semantical rules for S_2 are stated (§ 41). An analysis of the variables occurring in modal sentences shows that they have to be interpreted as referring to intensions (§ 40); hence a translation in words must be given either in terms of intensions (in the metalanguage M) or in neutral terms (in M′) (§ 43). Quine's views on the possibility of combining modalities and variables are discussed (§ 44). Finally, the main results of the discussions in this book are briefly summarized (§ 45).

§ 39. Logical Modalities

We form the modal system S_2 from our earlier system S_1 by the addition of the modal sign 'N' for logical necessity. We regard a proposition as necessary if any sentence expressing it is L-true. Other modalities can be defined in terms of necessity, for example, impossibility, possibility, contingency. With the help of 'N', we define symbols for necessary implication and necessary equivalence; the latter symbol may be regarded as an identity sign for intensions.

In the earlier chapters, modal sentences have sometimes been taken as examples, especially sentences about necessity or possibility, either in words (for instance, in §§ 30 and 31) or in symbols (for instance, § 11, Example II). We use 'N' as a sign for logical necessity; 'N(A)' is the symbolic notation for 'it is (logically) necessary that A'.

Quite a number of different systems of modal logic have been constructed, by C. I. Lewis (see Bibliography) and others.[1] These systems differ from one another in their basic assumptions concerning modalities. There is, for instance, the question of whether all sentences of the form '$Np \supset NNp$' are true, in words: 'if it is necessary that p, then it is necessary that it is necessary that p'. Some of the systems give an affirmative answer to this question, other systems give a negative answer or leave it undecided. Not only do logicians disagree among themselves on this question, but sometimes also one logician constructs systems which differ in this point, probably because he is doubtful whether he should regard the sentences mentioned as true or false. There are several further points of

[1] For bibliographical references up to 1938, see Church's bibliography in *Journal of Symbolic Logic*, Vols. I and III; the pertinent references are listed in III, 199 ("Modality") and 202 ("Strict Implication").

difference between the systems. All these differences are, I think, due to the fact that the concept of logical necessity is not sufficiently clear; it can, for instance, be conceived in such a way that the sentences mentioned are true, but also in another way such that they, or some of them, are false.

Our task will be to find clear and exact concepts to replace the vague concepts of the modalities as used in common language and in traditional logic. In other words, we are looking for explicata for the modalities. It seems to me that a simple and convenient way of explication consists in basing the modalities on the semantical L-concepts. The concept of logical necessity, as explicandum, seems to be commonly understood in such a way that it applies to a proposition *p* if and only if the truth of *p* is based on purely logical reasons and is not dependent upon the contingency of facts; in other words, if the assumption of not-*p* would lead to a logical contradiction, independent of facts. Thus we see a close similarity between two explicanda, the logical necessity of a proposition and the logical truth of a sentence. Now for the latter concept we possess an exact explicatum in the semantical concept of L-truth, defined on the basis of the concepts of state-description and range (2-2). Therefore, the most natural way seems to me to take as the explicatum for logical necessity that property of propositions which corresponds to the L-truth of sentences. Accordingly, we lay down the following convention for 'N':

39-1. For any sentence '. . .', 'N(. . .)' is true if and only if '. . .' is L-true.

We shall construct the system S_2 by adding to the system S_1 the sign 'N' with suitable rules such that the convention just stated is fulfilled (§ 41). This convention may be regarded as a rule of truth for the full sentences of 'N'. S_2 thus contains all the signs and the sentences of S_1.

On the basis of our interpretation of 'N', as given by the convention 39-1, the old controversies can be solved. Suppose that 'L-true in S_2' is defined in such a way that our earlier convention 2-1, which says that a sentence is L-true if and only if it is true in virtue of the semantical rules alone, independently of any extra-linguistic facts, is fulfilled. Let 'A' be an abbreviation for an L-true sentence in S_2 (for example, 'Hs $\vee \sim$ Hs'). Then 'N(A)' is true, according to 39-1. And, moreover, it is L-true, because its truth is established by the semantical rules which determine the truth and thereby the L-truth of 'A', together with the semantical rule for 'N', say 39-1. Thus, generally, if 'N(. . .)' is true, then 'NN(. . .)' is true; hence any sentence of the form 'N$p \supset$ NNp' is true. This constitutes an affirmative answer to the controversial question mentioned in the beginning. It can be shown in a similar way that every sentence of the form '$\sim \mathrm{N}p \supset \mathrm{N} \sim \mathrm{N}p$' is true. This settles another one of the controversial questions.[2]

This analysis leads to the result that, if 'N(. . .)' is true, it is L-true; and if it is false, it is L-false; hence:

39-2. Every sentence of the form 'N(. . .)' is L-determinate.

Therefore, the convention 39-1 may be replaced by the following more specific one:

39-3. For any sentence '. . .' in S_2, 'N(. . .)' is L-true if '. . .' is L-true; and otherwise 'N(. . .)' is L-false.

On the basis of the concept of logical necessity, the other logical modalities can easily be defined, as is well known. For example, 'p is impossible' means 'non-p is necessary'; 'p is contingent' means 'p is neither necessary nor impossible'; 'p is possible' means 'p is not impossible' (we adopt this interpretation in agreement with the majority of contemporary logicians, in distinction to other philosophers who use 'possible' in the sense of our 'contingent'). Let us use the diamond, '$\Diamond$', as a sign of possibility; we define it on the basis of 'N':

39-4. *Abbreviation.* '$\Diamond$(. . .)' for '$\sim \mathrm{N} \sim$(. . .)'.

It would also be possible to take '$\Diamond$' as primitive, as Lewis does, and then to define 'N(. . .)' by '$\sim \Diamond \sim$(. . .)'.

There are six modalities, that is, purely modal properties of propositions (as distinguished from mixed modal properties, for instance, contingent truth, see 30-1). The accompanying table shows how they can be

THE SIX MODALITIES

Modal Property of a Proposition	With 'N'	With '$\Diamond$'	Semantical Property of a Sentence
Necessary.	$\mathrm{N}p$	$\sim \Diamond \sim p$	L-true
Impossible.	$\mathrm{N} \sim p$	$\sim \Diamond p$	L-false
Contingent.	$\sim \mathrm{N}p \cdot \sim \mathrm{N} \sim p$	$\Diamond \sim p \cdot \Diamond p$	Factual
Non-necessary. . .	$\sim \mathrm{N}p$	$\Diamond \sim p$	Non-L-true
Possible.	$\sim \mathrm{N} \sim p$	$\Diamond p$	Non-L-false
Noncontingent. .	$\mathrm{N}p \vee \mathrm{N} \sim p$	$\sim \Diamond \sim p \vee \sim \Diamond p$	L-determinate

expressed in terms of 'N' and in terms of '$\Diamond$'. The last column gives the corresponding semantical concepts; a proposition has one of the modal

[2] The two questions and the reasons for our affirmative answers are discussed in more detail in [Modalities], § 1.

properties if and only if any sentence expressing the proposition has the corresponding semantical property.

Every proposition with respect to a given system *S* is either necessary or impossible or contingent. This classification is, according to our interpretation of the modalities, analogous to the classification of the sentences of *S* into the three classes of L-true, L-false, and factual sentences. There is, however, one important difference between the two classifications. The number of L-true sentences may be infinite, and it is, indeed, infinite for each of the systems discussed in this book. On the other hand, there is only one necessary proposition, because all L-true sentences are L-equivalent with one another and hence have the same intension. [This result holds only for that use of the term 'proposition' which is based on L-equivalence as the condition of identity. It is, of course, possible to choose a stronger requirement for identity, for instance, intensional isomorphism. In this case the intensional structures are called 'propositions'. And their number is infinite.] Likewise, there is only one impossible proposition, because all L-false sentences are L-equivalent. But the number of contingent propositions (with respect to a system with an infinite number of individuals) is infinite, like that of factual sentences.

It should be noted that the two sentences 'N(A)' and 'the sentence 'A' is L-true in S_2' correspond to each other merely in the sense that, if one of them is true, the other must also be true; in other words, they are L-equivalent (assuming that L-terms are defined in a suitable way so as to apply also to the metalanguage). This correspondence cannot be used as a *definition* for 'N', because the second sentence belongs, not to the object language S_2 as the first one does, but to the metalanguage M. The second sentence is not even a *translation* of the first in the strict sense which requires not only L-equivalence but intensional isomorphism (§ 14). If M contains the modal term 'necessary', then 'N(A)' can be translated into M by a sentence of the form 'it is necessary that . . .' (where '. . .' is the translation of 'A'). If M contains no modal terms, then there is no strict translation for 'N(A)'. But the correspondence stated makes it possible in any case to give an *interpretation* for 'N(A)' in M with the help of the concept of L-truth, for example, by laying down the truth-rule, 39-1.

On the basis of 'N', we introduce two further modal signs for modal relations between propositions:

39-5. *Abbreviation.* Let '. . .' and '- - -' be sentences in S_2. '. . . ⥽ - - -' for 'N(. . . ⊃ - - -)'.

39-6. *Abbreviation.* Let '. . .' and '- - -' be any designators in S_2 (sentences or otherwise). '. . . ≡ - - -' for 'N(. . . ≡ - - -)'.

Thus '⥽' is a sign for necessary implication between propositions (Lewis' strict implication). The symbol '≡' is a sign for necessary equivalence. The sign '≡' in S_2 is the analogue to the term 'L-equivalent' in its non-semantical use in M (5-4) or M′ (§ 34), where it designates a relation between intensions, not between designators. When standing between sentences, it corresponds to Lewis' sign '=' for strict equivalence. We have seen earlier that '≡', standing between designators of any type, is a sign for the identity of extensions (see remark on 5-3). Here in S_2, '≡' is, similarly, a sign for the identity of intensions. For example, 'H ≡ RA' is short for 'N(H ≡ RA)'. Hence, according to the rule 39-1, 'H ≡ RA' is true if and only if 'H ≡ RA' is L-true, hence if and only if 'H' and 'RA' are L-equivalent, in other words, have the same intension.

We have earlier formulated the two principles of interchangeability (12-1 and 12-2). For the first principle we have given, in addition to the chief formulation in semantical terms (12-1a), alternative formulations with the help of sentences of the object language containing '≡' (12-1b and c). Now, with the help of '≡', we can provide analogous formulations for the second principle. The following theorems 39-7b and c, which may be added to 12-2a as 12-2b and c, follow from 12-2a because $\mathfrak{A}_j$ and $\mathfrak{A}_k$ are L-equivalent if and only if $\mathfrak{A}_j \equiv \mathfrak{A}_k$ is true.

Second Principle of Interchangeability (alternative formulations):

39-7. Under the conditions of 12-2, the following holds:

b. (12-2b). $(\mathfrak{A}_j \equiv \mathfrak{A}_k) \supset (\ldots \mathfrak{A}_j \ldots \equiv \ldots \mathfrak{A}_k \ldots)$ is true (in *S*).

c. (12-2c). Suppose the system *S* contains variables for which $\mathfrak{A}_j$ and $\mathfrak{A}_k$ are substitutable, say '*u*' and '*v*'; then '$(u)(v)[(u \equiv v) \supset (\ldots u \ldots \equiv \ldots v \ldots)]$' is true (in *S*).

§ 40. Modalities and Variables

Problems concerning the interpretation of variables in modal sentences are discussed, in preparation for the semantical rules given in the next section. It is found that a universal quantifier preceding 'N' is to be interpreted as if it followed the 'N'. It is generally shown that variables in modal sentences are to be understood as referring to intensions rather than to extensions. Thus an individual variable in S_2 is interpreted as referring to individual concepts rather than to individuals. We decide to take as values of these variables not only those individual concepts which are expressible by descriptions in S_2 but the wider class of all individual concepts with respect to S_2. A concept of this kind is represented by any assignment of exactly one individual constant to each state-description in S_2.

So far we have given an interpretation for 'N' only in the case in which the argument-expression of 'N' is a sentence. But in a system which contains variables we also have to solve the problem of interpreting occurrences of 'N' followed by a matrix with free variables, e.g., 'N(P*x*)'. Let us investigate this problem in a general way for a system *S* containing a variable '*u*' of any type. How should we interpret the sentence '(*u*)[N (. . *u* . .)]', where '. . *u* . .' is a matrix containing '*u*' as the only free variable? Let us first consider the case in which '*u*' has only a finite number of values, say *n*, and all these values are expressible in *S*, say by the designators 'U_1', 'U_2', . . . 'U_n'. (As we shall see later, the interpretation of a variable in a modal sentence has to be given in terms of value-intensions, not value-extensions. Therefore, the statement just made is to be understood as saying that there are *n* value-intensions for '*u*' and that they are the intensions of the designators 'U_1', etc.) Now any universal sentence, whether in an extensional or in a modal language, always means that all values of the variable possess the property expressed by the matrix. Therefore, if the number of values is *n*, the universal sentence means the same as the conjunction of the *n* substitution instances of the matrix. In our example, '(*u*)[N(. . *u* . .)]' means the same as 'N(. . U_1 . .) • N(. . U_2 . .) • . . . • N(. . U_n . .)'.

A conjunction of *n* components ($n \geqq 2$) is L-true if and only if every one of the components is L-true. Therefore, the following holds, in virtue of the correspondence between necessity and L-truth (39-1):

40-1. If 'A_1', . . . 'A_n' are any sentences, 'N(A_1 • A_2 • . . . • A_n)' is L-equivalent to 'N(A_1) • N(A_2) • . . . • N(A_n)'.

If we apply this to the above result, we find that '(*u*)[N(. . *u* . .)]' means the same as 'N[(. . U_1 . .) • (. . U_2 . .) • . . . • (. . U_n . .)]' and hence the same as 'N[(*u*)(. . *u* . .)]'. Thus the result is that '(*u*)' and 'N' may exchange their places.

Next, let us consider the case in which the variable '*u*' has an infinite, but denumerable, number of values, all of which are expressible in *S*, say by the designators 'U_1', 'U_2', etc. Here we cannot form a conjunction of the substitution instances, but we can still consider their class. If we interpret a class of sentences as a joint assertion of its sentences, in accord with the usual procedure, then we can apply semantical concepts to it in the following way: We define the range of a class of sentences as the product of the ranges of the sentences. This leads to the following two results:

(i) A class of sentences is true if and only if all its sentences are true.
(ii) A class of sentences is L-true if and only if all its sentences are L-true.

Now the sentence '(*u*)[N(. . *u* . .)]' is true if and only if the class of the instances 'N(. . U_n . .)' for $n = 1, 2$, etc., is true; hence, according to (i), if and only if every sentence of the form 'N(. . U_n . .)' is true; hence, according to 39-1, if and only if every sentence of the form '. . U_n . .' is L-true; hence, according to (ii), if and only if the class of these sentences is L-true; hence, if and only if '(*u*) (. . *u* . .)' is L-true; hence, according to 39-1, if and only if 'N[(*u*) (. . *u* . .)]' is true. Thus the result is that, in the case of infinitely many values also, the quantifier '(*u*)' and the modal sign 'N' in the original sentence may exchange places.

It seems natural to apply the same result to the case in which not all values of '*u*' are expressible in *S*, that is to say, to interpret a sentence of the form '(*u*)[N(. . *u* . .)]' in any case, irrespective of the number and expressibility of the values of '*u*', as meaning the same as 'N[(*u*) (. . *u* . .)]'. In particular, we shall construct the semantical rules of the system S_2 in such a way that any two sentences of the forms just stated are L-equivalent (§ 41). In S_2 '*u*' must, of course, be an individual variable.

Since a modal system contains not only extensional but also intensional contexts, a designator may, in general, be replaced by another one only if they are not merely equivalent but L-equivalent. Thus, in general, we have to take into consideration the intensions of the designators, not merely their extensions. Similarly, we have to consider for a given variable its value-intensions in the first place. If the system contains variables of the type of sentences, say '*p*', '*q*', etc., then a quantifier with a variable of this kind occurring in a modal sentence must be interpreted as referring to propositions, not to truth-values. For example, the sentence '($\exists$*p*) ($\sim$N*p*)' must be understood as saying that there is a non-necessary proposition. It would hardly make sense to interpret it as saying that there is a non-necessary truth-value, because there are propositions with the same truth-value such that one of them fulfils the matrix '$\sim$N*p*', while another one does not. This interpretation in terms of propositions seems generally accepted. C. I. Lewis, as well as the other logicians who have discussed his systems of modal logic or have constructed new ones, have used interpretations in terms of propositions. If variables of the type of predicators of degree one occur in a modal system, it is clear that they must be interpreted analogously in terms of properties, not of classes. Here, again, I think that most logicians would agree;

however, modal sentences with variables of this kind have not been discussed frequently.

In my view the situation with respect to individual variables is quite analogous, although this is usually not recognized. I think that individual variables in modal sentences, for example, in S_2, must be interpreted as referring, not to individuals, but to individual concepts. The difficulties which would otherwise arise will be explained later (§ 43). Thus a sentence of the form '$(x)(\ldots x \ldots)$' in S_2 is to be interpreted as referring to all individual concepts. Therefore, we now have to study the question as to what is to be regarded as the totality of all individual concepts with respect to S_2.

We shall assume for the following discussions that the individual constants in S_2 are L-determinate (§ 19), that is to say that they are interpreted by the rules of designation as referring to positions in an ordered domain and that any two different constants refer to different positions. [For this purpose, it would be more natural to construct S_2 on the basis of S_3 (§ 18) rather than of S_1. The reason for taking S_1 as the basis is merely the possibility of using the earlier examples. But we must then suppose that, for example, the rule of designation for 's' does not use the phrase 'the man who was known by the name of 'Walter Scott' ', but rather: 'the man who was born at such and such a place at such and such a time'; and even this formulation would not be entirely adequate.] Consequently, we take any sentence of the form '$a \equiv b$' as L-false. However, $\equiv$-sentences with one or two descriptions (for example, '$(\imath x)(Axw) \equiv s$') are still, in general, factual.

A description $\mathfrak{A}_i$ in S_2, say '$(\imath x)(\ldots x \ldots)$', characterizes one of the individual positions with the help of the property expressed by the matrix '$\ldots x \ldots$'. If exactly one position has this property, then this position is the descriptum; otherwise, a^* is the descriptum (§ 8). Thus for the determination of the descriptum, the extension of $\mathfrak{A}_i$, factual investigation is required (unless the description is L-determinate). On the other hand, the intension of $\mathfrak{A}_i$, the individual concept expressed by $\mathfrak{A}_i$, must be something that can be determined by logical analysis alone. In order to understand more clearly what kind of entity an individual concept is, let us see what we can find out about the description $\mathfrak{A}_i$ by logical analysis alone. Suppose a state-description $\mathfrak{K}_n$ in S_2 is given (which is an infinite class of sentences in S_2). Then the question of whether or not there is exactly one individual position in $\mathfrak{K}_n$ fulfilling the matrix '$\ldots x \ldots$'—in other words, whether or not there is exactly one substitution instance of the matrix with an individual constant which holds in $\mathfrak{K}_n$—is a purely logical question. If the answer is in the affirmative, the descriptum of $\mathfrak{A}_i$ with respect to $\mathfrak{K}_n$ is represented by that one individual constant; otherwise it is represented by 'a*'. Thus the description $\mathfrak{A}_i$ assigns to every state-description exactly one individual constant; any individual constant may be assigned to several state-descriptions. If $\mathfrak{A}_i$ and $\mathfrak{A}_j$ are L-equivalent and hence express the same individual concept, then both assign to any state-description the same individual constant. Therefore, we might say that an individual concept with respect to S_2 is an assignment of exactly one individual to every state (which is a proposition expressed by a state-description). However, we shall actually take not these states but the state-descriptions; and not the individuals but the individual constants. The latter is possible because we have assumed that these constants are L-determinate and that there is a one-one correlation between the individuals and the individual constants. Thus we shall take any assignment of exactly one individual constant to each state-description in S_2 (in other words, any function from state-descriptions to individual constants) as representing an individual concept with respect to S_2. Only a small part (a denumerable class) of the individual concepts represented by assignments of this kind are expressible by descriptions in S_2. Now we decide to take as values of the individual variables in S_2 not only the individual concepts *expressible* by descriptions *in* S_2 but all individual concepts represented by assignments of the kind described; we call them individual concepts *with respect to* S_2. In the next section we shall lay down the semantical rules for S_2 in accord with this decision; a universal quantifier will be interpreted as referring to all individual concepts with respect to S_2.

Some remarks may, incidentally, be made concerning the interpretation of variables of other than individual type. Let *S* be a modal system which also contains propositional variables '*p*', etc., and variables '*f*', etc., for properties of level one, that is, properties of individuals. As values for propositional variables we should take not only those propositions which are expressed by sentences in *S*, but all propositions with respect to *S*. They are represented by the ranges in *S*, that is, the classes of state-descriptions in *S*. And as values for '*f*', etc., we should take not only those properties which are expressed by predicators (including lambda-expressions) in *S*, but all properties with respect to *S*. Since the attribution of a property to an individual results in a proposition, we may regard a property as an assignment of exactly one proposition to each individual. Therefore, we may represent the properties with respect to *S* by the assignments of ranges (classes of state-descriptions) in *S* to the individual constants in *S*. Similarly, assignments of ranges in *S* to ordered pairs of individual constants in *S* may be taken as representing the relations with respect to *S* as values of relation variables in *S*. [In analogy to the rules of ranges for matrices containing individual variables in S_2, which will be given in the next section, rules for variables of other types in *S* might be stated as follows: (i) The matrix '*p*' holds in the state-description $\mathfrak{R}_n$ for a certain range as value if and only if $\mathfrak{R}_n$ belongs to this range. (ii) The matrix '*f*a' holds in $\mathfrak{R}_n$ for a given assignment of the kind described as value of '*f*' if and only if $\mathfrak{R}_n$ belongs to that range which is assigned to 'a'.]

§ 41. Semantical Rules for the Modal System S_2

On the basis of our previous decisions concerning the interpretation of 'N' (§ 39) and of the individual variables in S_2 (§ 40), we lay down semantical rules for S_2. The most important rules are the rules of ranges, which are here somewhat more complicated than for S_1 because individual concepts rather than individuals must here be taken as values of the variables. The L-concepts for S_2 have the same definitions as for S_1. Some examples of L-true modal sentences in S_2 are given.

The signs of the modal system S_2 comprise those of S_1 and, in addition, the modal sign 'N'. In S_1, compound designators and designator matrices are formed out of atomic matrices with the help of the following means: the ordinary (i.e., nonmodal) connectives, quantifiers, the iota-operator, and the lambda-operator. In S_2 a rule of formation for 'N' is added, which says that, if '. . .' is any matrix, 'N(. . .)' is a matrix.

Now we have to construct the rules of ranges for S_2. The state-descriptions in S_2 are the same as in S_1 (§ 2), because S_2 does not contain any new descriptive constants. If we had only sentences without variables, we could simply take the rules of ranges for S_1 (see the examples in § 2, omitting the rule for a universal sentence) and add the following rule:

41-1. N($\mathfrak{S}_i$) holds in every state-description if $\mathfrak{S}_i$ holds in every state-description; otherwise, N($\mathfrak{S}_i$) holds in no state-description.

This rule is clearly in accord with our convention 39-3 (see 2-2 and 2-4). However, in order to accommodate sentences with variables, we have to use, instead, more complicated rules of ranges. They must apply not only to sentences, like the rules of ranges for S_1 (§ 2), but to matrices, and they must refer to values of the individual variables occurring in the matrix. According to our analysis in the preceding section, we take as values of the variables all individual concepts with respect to S_2; every one of these concepts is represented by an assignment of individual constants to state-descriptions. Suppose that we have chosen as a value of the variable 'x' occurring in the atomic matrix 'Px' an assignment of this kind and that the individual constant assigned to a given state-description $\mathfrak{R}_n$ is 'b'. Then the question of whether the matrix 'Px' for the chosen value of 'x' holds in $\mathfrak{R}_n$ means simply whether the sentence 'Pb' holds in $\mathfrak{R}_n$; and this is, of course, the case if 'Pb' belongs to $\mathfrak{R}_n$ (compare the example (1) of the rules of ranges for S_1 in § 2). This analysis suggests the first of the subsequent rules of ranges (41-2a). The other rules are analogous to the rules of ranges for S_1 (§ 2), together with the rule 41-1 for 'N', except that the present rules apply to matrices and therefore have to refer to assignments as values of the free variables.[3] Note that sentences are matrices without free variables (§ 1); therefore, these rules apply also to sentences, in which case the references to values are dropped.

41-2. ***Rules of ranges*** for the modal system S_2. Let $\mathfrak{A}_i$ be a matrix and $\mathfrak{R}_n$ be a state-description in S_2. By a value of a variable we mean any assignment of the kind described earlier.

a. Let $\mathfrak{A}_i$ be of atomic form. $\mathfrak{A}_i$ holds in $\mathfrak{R}_n$ for given values of the individual variables occurring in $\mathfrak{A}_i$, if and only if $\mathfrak{R}_n$ contains the atomic sentence formed from $\mathfrak{A}_i$ by substituting for every free variable the constant assigned to $\mathfrak{R}_n$ by the value of the variable.

b. Let $\mathfrak{A}_i$ be an $\equiv$-matrix with individual signs (constants or variables). $\mathfrak{A}_i$ holds in $\mathfrak{R}_n$ for given values of the variables occurring in $\mathfrak{A}_i$, if the individual constant for the left side (that is, either

[3] The system MFL described in [Modalities], § 9, is similar to, but somewhat simpler than, our present system S_2. Sentences of the form 'a = b' in MFL are regarded as L-false, like the corresponding sentences of the form 'a $\equiv$ b' in S_2; this shows that the individual constants in MFL are, in terms of our present theory, L-determinate like those in S_2. The state-descriptions are the same in both systems. The differences are as follows: MFL does not contain lambda-expressions and individual descriptions; this difference is not essential, since both kinds of expressions in S_2 can be eliminated, as we have seen. More essential is the difference in the interpretation of individual variables. A universal sentence '$(x)(\ldots x \ldots)$' in MFL is regarded as L-equivalent to the class of substitution instances of the matrix '$\ldots x \ldots$' with all individual constants; thus, in terms of our present theory, the universal quantifier refers to all L-determinate individual concepts and to no others. A universal quantifier in S_2, on the other hand, refers to all individual concepts (with respect to S_2). This wider range of values for the individual variables in S_2 seems more adequate; but it makes necessary the somewhat more complicated form of the rules of ranges as given in the text, while the rules of ranges for MFL are as simple as those for S_1, together with the rule 41-1 for 'N'.

the individual constant standing on the left side or the individual constant assigned to $\mathfrak{R}_n$ by the value of the variable standing on the left side) is the same as that for the right side.

c. Let $\mathfrak{A}_i$ be $\sim\mathfrak{A}_j$. $\mathfrak{A}_i$ holds in $\mathfrak{R}_n$ for given values of the variables occurring freely in $\mathfrak{A}_i$, if $\mathfrak{A}_j$ does not hold in $\mathfrak{R}_n$ for these values.

d. Let $\mathfrak{A}_i$ be $\mathfrak{A}_j \vee \mathfrak{A}_k$. $\mathfrak{A}_i$ holds in $\mathfrak{R}_n$ for given values of the free variables, if either $\mathfrak{A}_j$ or $\mathfrak{A}_k$ or both hold in $\mathfrak{R}_n$ for these values.

e. Let $\mathfrak{A}_i$ be $\mathfrak{A}_j \bullet \mathfrak{A}_k$. $\mathfrak{A}_i$ holds in $\mathfrak{R}_n$ for given values of the free variables, if both $\mathfrak{A}_i$ and $\mathfrak{A}_k$ hold in $\mathfrak{R}_n$ for these values.

f. Let $\mathfrak{A}_i$ consist of a universal quantifier followed by the matrix $\mathfrak{A}_j$ as its scope. $\mathfrak{A}_i$ holds in $\mathfrak{R}_n$ for given values of the variables occurring freely in $\mathfrak{A}_i$ (hence not including the variable occurring in the initial quantifier), if $\mathfrak{A}_j$ holds in $\mathfrak{R}_n$ for every value of the variable of the initial quantifier and the given values of the other free variables.

g. Let $\mathfrak{A}_i$ be $\mathrm{N}(\mathfrak{A}_j)$. $\mathfrak{A}_i$ holds in $\mathfrak{R}_n$ for given values of the free variables, if $\mathfrak{A}_j$ holds in every state-description for these values.

The following two theorems are simple consequences of these rules; they may be used instead of the rules for the determination of the range of a nonmodal matrix or sentence in S_2.

41-3. Let $\mathfrak{A}_i$ be a matrix of any form without 'N' in S_2. $\mathfrak{A}_i$ holds in $\mathfrak{R}_n$ for given values of the free variables, if and only if the sentence formed from $\mathfrak{A}_i$ by substituting for every free variable the constant assigned to $\mathfrak{R}_n$ by the value of the variable holds in $\mathfrak{R}_n$.

41-4. If a sentence in S_2 does not contain 'N', then it holds in S_2 in the same state-descriptions as in S_1.

In order to avoid certain complications, which cannot be explained here, it seems advisable to admit in S_2 only descriptions which do not contain 'N'. But any description may, of course, occur within the scope of an 'N'. The smallest matrix in which a description occurs (in the primitive notation) is always a nonmodal context, because the description must be an argument expression either of a primitive predicator constant or of '$\equiv$'. This smallest matrix is then taken as the context '- -$(\imath x)$ $(\,.\,.\ x\ .\,.)$ - -', which can be transformed into 8-2. In this way every description can be eliminated. Since L-equivalent sentences are L-interchangeable also within modal contexts, according to the second principle of interchangeability (12-2), the result of the elimination is L-equivalent to the original sentence; or, rather, we lay down a rule to the effect that any sentence containing descriptions holds in the same state-descriptions as the sentence resulting from the described elimination of the descriptions, and hence the two sentences become L-equivalent.

Another point is worth noting. Although we interpret the individual variables in S_2 as referring to individual concepts, not to individuals, nevertheless a description in S_2 characterizes, not one individual concept, but mutually equivalent individual concepts—in other words, one individual. This follows from the rule just mentioned, which permits the transformation into 8-2. The first part of 8-2 says, in words: 'there is an individual concept *y* such that, for every individual concept *x*, *x* has the descriptional property if and only if *x* is *equivalent* (not 'L-equivalent' or 'identical'!) to *y*'; in other words, 'all individual concepts equivalent to *y*, and only these, have the property'; or, 'the *individual* *y* is the only individual which has the property'. This is as it should be, because the purpose of a description, even in a modal language, is to refer to one individual with the help of a property possessed by that individual alone. Nevertheless, the description has, of course, a unique intension, which is an individual concept. This individual concept is not the only one possessing the descriptional property, since, as we have seen, all equivalent ones do likewise; but it is uniquely determined by the descriptional property; as Frege puts it, it is not the individual but the way in which the description refers to the individual.

For lambda-expressions we do not impose the restriction stated for descriptions; they may also contain 'N'. Any lambda-operator can be eliminated in S_2 by conversion in the same way as in S_1 (§ 1). Here, again, a rule would be laid down saying that a sentence containing lambda-operators holds in the same state-descriptions as the sentence resulting from their elimination.

The L-concepts are defined for S_2 in the same way as for S_1 (§ 2). The following theorems give a few results, which hold on the basis of the rules of ranges stated above.

41-5. Any sentence of one of the following forms is L-true in S_2. (The variables '*p*', '*q*', . . '*f*', do not occur in S_2 but are here used merely to describe forms of sentences in S_2. A sentence in S_2 is said to have one of the forms described if it is formed by substituting for '*p*' or '*q*' any sentence in S_2 and for '*fx*' any matrix containing '*x*' as the only free variable.)

a. '$\mathrm{N}p \supset p$'.
b. '$p \supset \Diamond p$'.
c. '$(p \supset q) \supset (\mathrm{N}p \supset \mathrm{N}q)$'
d. '$\mathrm{N}(p \bullet q) \equiv \mathrm{N}p \bullet \mathrm{N}q$'.
e. '$\Diamond(p \vee q) \equiv \Diamond p \vee \Diamond q$'.
f. '$\mathrm{NN}p \equiv \mathrm{N}p$'.
g. '$\mathrm{N} \sim \mathrm{N}p \equiv \sim \mathrm{N}p$'.
h. '$\Diamond\Diamond p \equiv \Diamond p$'.
i. '$\Diamond \mathrm{N}p \equiv \mathrm{N}p$'.
j. '$\mathrm{N}\Diamond p \equiv \Diamond p$'.
k. '$(x)\mathrm{N}(fx) \equiv \mathrm{N}(x)(fx)$'.
l. '$(\exists x)\mathrm{N}(fx) \supset \mathrm{N}(\exists x)(fx)$'.
m. '$(\exists x)\Diamond(fx) \equiv \Diamond(\exists x)(fx)$'.
n. '$\Diamond(x)(fx) \supset (x)\Diamond(fx)$'.

We see from these theorems that 'N' is quite similar to a universal quantifier and '$\Diamond$' to an existential quantifier. This seems plausible, since $\mathrm{N}\mathfrak{S}_i$ is true if $\mathfrak{S}_i$ holds in every state-description, and $\Diamond\mathfrak{S}_i$ is true if $\mathfrak{S}_i$ holds in at least one state-description.

§ 42. Modalities in the Word Language

The problem of the translation of modal sentences of S_2 into the metalanguages M and M′ is discussed. It is shown that it is advisable to use for the translations either terms of intensions in M or neutral terms in M′. The use of terms of extensions within modal sentences in M is not in itself incorrect, provided that certain restrictions are observed; but it involves the danger of making wrong inferences by overlooking the restrictions.

We shall examine here the problem of the formulation of modal sentences in words and, in particular, the problem of the translation of modal sentences into our metalanguages M and M′. It is worth while to study this problem because, it seems to me, certain difficulties which have sometimes been found in connection with modal sentences are due chiefly to their inadequate or misleading formulation in the word language.

Since modal sentences, for instance, in S_2 or in a richer language with several types of variables, are not semantical, their translations are likewise not semantical sentences and hence belong to the nonsemantical part of M and M′ (this part of M′ was explained in §§ 34–36). As translation of 'N', we take 'it is necessary that'; hence, this is an intensional phrase.

We shall discuss three examples—A, B, and C. In A, we have predicators as argument expressions of '$\equiv$' or '$\equiv$'; in B, sentences; in C, individual expressions. Otherwise, the three examples are perfectly analogous. Therefore, we arrange them in three parallel columns. This facilitates the comparison of corresponding expressions in the three examples and the recognition of their analogy.

Because of the perfect analogy, any one of the three examples would theoretically be sufficient. However, for practical reasons it seems advisable to give all three. The purpose of the analysis of the examples is to show that it is advisable to formulate modal sentences either in terms of intensions or in neutral terms, while formulation in terms of extensions involves certain dangers. Now this result is easily seen in the case of predicators; presumably, most readers will agree in this case. Then the analogy will make it easier to recognize the same situation in the case of sentences and, finally, in the case of individual expressions. In this last case the inhibitions against a translation in terms of intensions are strongest because it is not customary to speak of individual concepts. Therefore, here the help of the two other examples seems necessary for practical, psychological reasons, although theoretically the situation is here as clear and simple as in the first two cases.

The example A (the conjunction of 42-1A and 42-2aA) is similar to one given by Church;[4] our '∼N(. . .)' corresponds to his '◇∼(. . .)'. In the example C, we use 'au' as abbreviation for '(ɿ*x*) (A*x*w)'. In the translation of this description into the word language, we omit, for the sake of brevity, the phrase 'or a*, if there is not exactly one such individual' (as we did earlier, § 9).

The following sentences in S_2 are true but not L-true (see 3-7 and 9-2):

42-1.

A	B	C
'F•B ≡ H'.	'(F•B)s ≡ Hs'.	'au ≡ s'.

Therefore, according to 39-1, prefixing 'N' yields false sentences; hence the following is true:

42-2a.

A	B	C
'∼N(F•B ≡ H)';	'∼N[(F•B)s ≡ Hs]';	'∼N(au ≡ s)';

or, abbreviated with '≢' (39-6):

42-2b.

A	B	C
'∼(F•B ≡ H)'.	'∼[(F•B)s ≡ Hs]'.	'∼(au ≡ s)'.

Now let us examine the question of the translations of these sentences of S_2 into M. The first sentence, 42-1 (in each of the three examples), is a nonmodal sentence. It can be translated in two different ways, either into 42-3 in terms of intensions with the nonsemantical term 'equivalent' (see 5-3 and 5-5) or into 42-4 in terms of extensions with the identity phrase 'is the same as' (see 4-7 and 9-1):

[4] [Review Q.], p. 46.

42-3.

A	B	C
'The property Featherless Biped is equivalent to the property Human'.	'The proposition that Scott is a featherless biped is equivalent to the proposition that Scott is human'.	'The individual concept The Author Of Waverley is equivalent to the individual concept Walter Scott'.

42-4.

A	B	C
'The class Featherless Biped is the same as the class Human'.	'The truth-value that Scott is a featherless biped is the same as the truth-value that Scott is human'.	'The individual The Author Of Waverley is the same as the individual Walter Scott'.

For the modal sentences 42-2, however, the situation is different. First, we shall give the translation into M in terms of intensions. We base the translation 42-5 on the second of the two notations a and b given for 42-2, utilizing the fact that '≡' is a sign for the identity of intensions (§ 39). (For A, see 4-8; for B, 6-4; for C, § 9).

42-5.

A	B	C
'The property Featherless Biped is not the same as the property Human'.	'The proposition that Scott is a featherless biped is not the same as the proposition that Scott is human'.	'The individual concept The Author Of Waverley is not the same as the individual concept Walter Scott'.

This translation is adequate and unobjectionable. Not so, however, the following translation in terms of extensions; here we base the translation on the first notation 42-2a and regard '≡' as a sign for the identity of extensions (see remark on 5-3).

42-6.

A	B	C
'It is not necessary that the class Featherless Biped is the same as the class Human'.	'It is not necessary that the truth-value that Scott is a featherless biped is the same as the truth-value that Scott is human'.	'It is not necessary that the individual The Author Of Waverley is the same as the individual Walter Scott'.

Formulations of this kind might perhaps be admitted as sentences in M; if so, they would presumably be regarded as true and as correct translations of 42-2a. However, these formulations are dangerous; if we apply customary ways of thinking to them, we obtain false results. In the ordinary word language, we are accustomed to using the principle of interchangeability (24-3b) implicitly. If in any of the three examples we apply this principle to 42-6 on the basis of the true identity sentence 42-4, we obtain the following result, 42-7. This, however, if admitted at all as a sentence, will certainly be regarded as false.

42-7.

A	B	C
'It is not necessary that the class Human is the same as the class Human'.	'It is not necessary that the truth-value that Scott is human is the same as the truth-value that Scott is human'.	'It is not necessary that the individual Walter Scott is the same as the individual Walter Scott'.

These are instances of the antinomy of the name-relation in its second form, similar to our previous example (§ 31). In spite of this result, we may admit the formulations 42-6, provided that we are willing to prohibit the use of the principle of interchangeability in cases of nonextensional contexts. However, since the unrestricted use of this principle is customary and plausible, there would always be the danger of forgetting the prohibiting rule and using the principle inadvertently. Therefore, it seems more advisable to avoid formulations like 42-6 and, in general, formulations in terms of extensions within modal or other nonextensional contexts.

Now let us see how the given symbolic sentences of S_2 are to be translated into the neutral metalanguage M′. As explained earlier, there are no identity phrases in M′; instead, the terms 'equivalent' and 'L-equivalent' are applied in their nonsemantical use (see 34-8 and 34-9). As 'equivalent' is a direct translation of the symbol '≡', so is 'L-equivalent' of '≣'. (This shows again that the nonsemantical term 'L-equivalent' is intensional; this holds for all nonsemantical (absolute) L-terms, see [I], § 17.) Thus the translation of 42-1 into M′ is as follows (see 34-10 and 34-13):

42-8.	A	B	C
	'Featherless Biped is equivalent to Human'.	**a.** 'That Scott is a featherless biped, is equivalent to that Scott is human'.	'The Author Of Waverley is equivalent to Walter Scott'.
		b. 'Scott is a featherless biped if and only if Scott is human'.	

In B we add here the alternative form b because it sounds more natural (see end of § 34).

There are two ways of translating 42-2 into M′. The first is based on 42-2a and translates 'N' by 'it is necessary that'. (In B we use again the more natural phrase 'if and only if' instead of 'is equivalent to'; concerning the reason for the word order, see remark at the end of § 34.)

42-9a.	A	B	C
	'It is not necessary that Featherless Biped is equivalent to Human'.	'That Scott is a featherless biped if and only if Scott is human, is not necessary'.	'It is not necessary that The Author Of Waverley is equivalent to Walter Scott'.

The second alternative is based on the notation 42-2b and translates '≡' by 'L-equivalent' (see 34-11):

42-9b.	A	B	C
	'Featherless Biped is not L-equivalent to Human'.	'That Scott is a featherless biped, is not L-equivalent to that Scott is human'.	'The Author Of Waverley is not L-equivalent to Walter Scott'.

This translation does not involve any difficulty analogous to that connected with 42-6.

Thus the final result is as follows: It seems advisable to frame the formulation of modal and other nonextensional sentences in the word language, not in terms of extensions, but either (i) in terms of intensions or (ii) in neutral terms. Which of the two formulations (i) and (ii) one prefers is a matter of practical decision (see the discussion at the end of § 37). The formulation in neutral terms is simpler, but the nonsemantical use of the terms 'equivalent' and 'L-equivalent' is not customary. Formulations in terms of intensions, like 42-5, are, in general, more customary, except for the reference to individual concepts in case C. But this reference will perhaps appear less strange if we recognize the essential analogy in 42-5 between C, on the one hand, and A and B, on the other.

§ 43. Modalities and Variables in the Word Language

Translations of symbolic modal sentences with variables into M and M′ are examined. The result is analogous to that in the preceding section. It is advisable to avoid terms of extensions and to use either terms of intensions in M or the neutral terms in M′. The translation in terms of propositions and properties is customary, but that in terms of individual concepts instead of individuals may at first appear strange.

We have seen earlier (§ 10) that, as a designator has both an extension and an intension, a variable has both value-extensions and value-intensions. Therefore, a sentence with a variable can be translated into M either in terms of its value-extensions or in terms of its value-intensions. Furthermore, it can be translated into M′ in neutral terms (§ 36). In analogy to the result in the preceding section, we shall find here that it is advisable to avoid the formulation in terms of value-extensions and to use either terms of value-intensions or neutral terms.

For the same reason as in the preceding section, we use here three analogous examples, A, B, and C. They are existential sentences with the variables '*f*', '*p*', and '*x*' in a modal system *S* containing variables of these types and the modal sign 'N'.

The following sentences 43-1a and b differ only in their notation. In each of the three examples, A, B, and C, 43-1a is derived by existential generalization from the conjunction of the sentences 42-1 and 42-2a; and likewise 43-1b from 42-1 and 42-2b.

43-1a.

A	B	C
'$(\exists f)[(f \equiv \mathrm{H}) \bullet \sim\mathrm{N}(f \equiv \mathrm{H})]$'.	'$(\exists p)[(p \equiv \mathrm{Hs}) \bullet \sim\mathrm{N}(p \equiv \mathrm{Hs})]$'.	'$(\exists x)[(x \equiv \mathrm{s}) \bullet \sim\mathrm{N}(x \equiv \mathrm{s})]$'.

43-1b.

A	B	C
'$(\exists f)[(f \equiv \mathrm{H}) \bullet \sim(f \equiv \mathrm{H})]$'.	'$(\exists p)[(p \equiv \mathrm{Hs}) \bullet \sim(p \equiv \mathrm{Hs})]$'.	'$(\exists x)[(x \equiv \mathrm{s}) \bullet \sim(x \equiv \mathrm{s})]$'.

We shall now examine the possibilities for the translation of these sentences into M. If it were a question of an extensional existential sentence —for instance, 43-1a with the second conjunctive component omitted— then translations in terms of value-intensions and of value-extensions would be equally acceptable. This, however, is not the case for these modal sentences. We shall first give a translation in terms of value-intensions, in analogy to 42-3 and 42-5, taking notation 43-1b and translating '$\equiv$' by identity of intensions:

43-2.

A	B	C
'There is a property f which is equivalent to but not the same as the property Human'.	'There is a proposition p which is equivalent to but not the same as the proposition that Scott is human'.	'There is an individual concept x which is equivalent to but not the same as the individual concept Walter Scott'.

In each of the three examples, this sentence can be derived by existential generalization from the conjunction of 42-3 and 42-5.

Now we shall translate 43-1a in terms of value-extensions, in analogy to 42-4 and 42-6, translating '$\equiv$' by identity of extensions:

43-3.

A	B	C
'There is a class f which is the same but not necessarily the same as the class Human'.	'There is a truth-value p which is the same but not necessarily the same as the truth-value that Scott is human'.	'There is an individual x which is the same but not necessarily the same as the individual Walter Scott'.

In each of the three examples, this sentence can be derived by existential generalization from the conjunction of 42-4 and 42-6. We have seen in the preceding section that formulations of modal sentences in terms of extensions, like 42-6, are dangerous because they lead to the antinomy of the name-relation unless special restrictions are imposed and that it is therefore advisable to avoid these formulations. The same holds for formulations like 43-3.

The translation of 43-1 into neutral formulations in M′, in analogy to 42-8 and 42-9b, is as follows:

43-4.

A	B	C
'There is an f such that f is equivalent but not L-equivalent to Human'.	'There is a p such that p is equivalent but not L-equivalent to that Scott is human'.	'There is an x such that x is equivalent but not L-equivalent to Walter Scott'.

(Use of 'F-equivalent' as a nonsemantical term would provide a shorter formulation.) In each of the three examples this sentence can be derived by existential generalization from the conjunction of 42-8 and 42-9b. The formulations 43-4 are free of the dangers involved in 43-3.

Now let us compare the three examples, A, B, and C. Our proposal not to translate variables in modal sentences in terms of extensions seems quite natural in cases B and A. As remarked earlier (§ 40), it seems that all logicians interpret modal sentences in terms of propositions rather than of truth-values, and most of them use terms of properties rather than of classes. Only in case C does our interpretation deviate from the customary one. The reference to individual concepts may first appear somewhat strange; and the alternative translation in neutral terms (e.g., 43-4C), which avoids the reference to individual concepts, uses the unfamiliar terms 'equivalent' and 'L-equivalent'. However, I believe that, once we are aware of the perfect analogy between the three cases, we recognize the inadequacy of the formulations in terms of individuals; and the impression of strangeness which the formulation in terms of individual concepts and, to a lesser degree, the neutral formulation may first give will perhaps disappear. Modal sentences with variables are of a quite peculiar logical nature, and it should not be surprising that an adequate and correct rendering for them in the word language is not always possible in entirely customary and natural terms.

§ 44. Quine on Modalities

Quine's article [Notes] explained his view that, under customary conditions, modalities and quantification cannot be combined. A new statement by Quine is quoted here, in which he says that my language succeeds in combining modalities with quantification but only at the price of repudiating all extensions, for instance, classes and individuals. I try to show that my modal language does not exclude anything that is admitted by a corresponding extensional language.

Quine[5] illustrates the difficulty which we have called the antinomy of the name-relation by the following example among others (as mentioned above, § 31). We find as an arithmetical and hence logical truth:

(i) '9 is necessarily greater than 7'.

The following is a true statement of astronomy:

(ii) 'The number of planets = 9'.

[5] Quine [Notes] (18) p. 121, (15) p. 119, (23) p. 121.

If, in (i), '9' is replaced by 'the number of planets' in virtue of the true identity statement (ii), we obtain the false statement:

(iii) 'The number of planets is necessarily greater than 7'.

Quine's method for solving the antinomy has been explained earlier (§ 32, Method II). According to our method, the following sentence takes the place of (ii) in M′:

(iv) 'The number of planets is equivalent to 9'.

The sentences (i) and (iii) occur also in M′. But now it is not possible to infer the false sentence (iii) from the true sentence (i) together with (iv). According to the first principle of interchangeability (12-1), the expressions 'the number of planets' and '9' are interchangeable on the basis of (iv) in extensional contexts only, hence not in (i). Thus the difficulty disappears, and the designators occurring in nonextensional contexts still function, according to our conception, as normal designators.

An even more serious problem is raised by Quine's objection to modal sentences with variables. He discusses the following expression:

(v) 'There is something which is necessarily greater than 7'.

He says[6] that this expression "is meaningless. For, would 9, that is, the number of planets, be one of the numbers necessarily greater than 7? But such an affirmation would be at once true in the form . . . [our (i)] and false in the form . . . [our (iii)]." Quine does not regard (i) and (iii) as meaningless. As explained earlier (§ 32, Method II), he regards occurrences of designators in nonextensional contexts, e.g., '9' in (i) and 'the number of planets' in (iii), as "not purely designative"; in other words, these occurrences do not function as names, and hence the principle of interchangeability is not applicable. For the same reason, according to Quine's view, the rule of existential generalization is not applicable to these occurrences. Therefore, there is no valid inference from (i) to (v), and, moreover, (v) has no meaning and hence cannot be admitted as a sentence. Thus Quine arrives at the following conclusions, which are stated at the end of his paper: "A substantive word or phrase which designates an object may occur purely designatively in some contexts and not purely designatively in others. This second type of context, though not less "correct" than the first, is not subject to the law of substitutivity of identity nor to the laws of application and existential generalization. Moreover, no pronoun (or variable of quantification) within a context of this

[6] *Ibid.*, p. 124.

second type can refer back to an antecedent (or quantifier) prior to that context. This circumstance imposes serious restrictions, commonly unheeded, upon the significant use of modal operators, as well as challenging that philosophy of mathematics which assumes as basic a theory of attributes [i.e., properties] in a sense distinct from classes."[7]

To Quine's contexts of the second kind belong all those which we call nonextensional. He discusses, in particular, contexts within quotes and modal contexts. With respect to contexts within quotes his conclusions are no doubt correct. I cannot agree, however, with Quine's conclusion concerning modal contexts. We have combined modalities and variables both in symbolic object languages (§ 40) and in word formulations in our metalanguages (§ 43).

Church likewise does not accept Quine's result. He says in the review of Quine's paper that he "would question strongly the conclusion which the author draws that no variable within an intensional context . . . can refer back to a quantifier prior to that context . . . . The conclusion should rather be that in order to do this a variable must have an intensional range—a range, for instance, composed of attributes [properties] rather than classes."[8] Up to this point I am in agreement with Church. His solution is as follows: He distinguishes, like the system PM (see § 27), between class variables, e.g., 'α', and property variables, e.g., 'ϕ'. He takes as example a sentence which is essentially the same as a conjunction of 42-1A and 42-2aA. In distinction to Quine, he regards it as admissible to infer from this sentence by existential generalization an existential sentence; the latter, however, must not have the form '$(\exists\alpha)(\ldots\alpha\ldots)$' but rather the form '$(\exists\phi)(\ldots\phi\ldots)$'. It seems to me that this procedure is correct and, indeed, solves completely the difficulty pointed out by Quine. I believe, however, that there is a simpler way to achieve this. It is similar to that of Church but avoids the use of two kinds of variables for the same type. This use is, as explained earlier (§ 27), an unnecessary duplication. It is sufficient to use variables of one kind which are neutral in the sense that they have classes as value-extensions and properties as value-intensions; this is done in 43-1aA. The use of different variables for extensions and intensions within all types would lead in the case of Quine's example (v) to the introduction of variables for number concepts different from the variables for numbers. This, however, would be both unnecessary and unusual.

The problem of whether or not it is possible to combine modalities and

[7] *Ibid.*, p. 127.

[8] [Review Q.], p. 46.

variables in such a way that the customary inferences of the logic of quantification—in particular, specification and existential generalization—remain valid is, of course, of greatest importance. Any system of modal logic without quantification is of interest only as a basis for a wider system including quantification. If such a wider system were found to be impossible, logicians would probably abandon modal logic entirely. Therefore, it is essential to clarify the situation created by Quine's analysis and objections. For this reason I have asked Quine, who has read an earlier version of the manuscript of this book, for a statement of his present view on the problem mentioned and, in particular, his reaction to my method for combining modalities and variables as explained in the preceding section. With his kind permission, I am quoting here his statement in full:[9]

> Every language system, insofar at least as it uses quantifiers, assumes one or another realm of entities which it talks about. The determination of this realm is not contingent upon varying metalinguistic usage of the term 'designation' or 'denotation', since the entities are simply the values of the variables of quantification. This is evident from the meaning of the quantifiers '(x)', '(f)', '(p)', '$(\exists x)$', '$(\exists f)$', '$(\exists p)$' themselves: 'Every (or, Some) entity x (or f or p) is such that'. The question *what there is* from the point of view of a given language —the question of the *ontology* of the language—is the question of the range of values of its variables.
>
> Usually the question will turn out to be in part an a priori question regarding the nature and intended interpretation of the language itself, and in part an empirical question about the world. The general question whether for example individuals, or classes, or properties, etc., are admitted among the values of the variables of a given language, will be an a priori question regarding the nature and intended interpretation of the language itself. On the other hand, supposing individuals admitted among the values, the further question whether the values comprise any unicorns will be empirical. It is the former type of inquiry—ontology in a philosophical rather than empirical sense—that interests me here. Let us turn our attention to the ontology, in this sense, of your object language.
>
> An apparent complication confronts us in the so-called duality of M′ as between intensional and extensional values of variables; for it would appear then that we must inquire into two alternative ontologies of the object language. This, however, I consider to be illusory; since the duality in question is a peculiarity only of a special metalinguistic idiom and not of the object language itself, there is nothing to prevent our examining the object language from the old point of view and asking what the values of its variables are in the old-fashioned non-dual sense of the term.
>
> It is now readily seen that those values are merely intensions, rather than extensions or both. For, we have:
>
> $(x)(x \equiv x)$,
>
> i.e., every entity is L-equivalent to itself. This is the same as saying that entities between which L-equivalence fails are distinct entities—a

[9] The first two-thirds of Quine's statement as here quoted is dated October 23, 1945; the remainder January 1, 1946.

clear indication that the *values* (in the ordinary non-dual sense of the term) of the variables are properties rather than classes, propositions rather than truth-values, individual concepts rather than individuals. (I neglect the further possibility of distinctness among L-equivalent entities themselves, which would compel the entities to be somehow "ultra-intensional"; for it is evident that you have no cause in the present connection to go so far.)

I agree that such adherence to an intensional ontology, with extrusion of extensional entities altogether from the range of values of the variables, is indeed an effective way of reconciling quantification and modality. The cases of conflict between quantification and modality depend on extensions as values of variables. In your object language we may unhesitatingly quantify modalities because extensions have been dropped from among the values of the variables; even the individuals of the concrete world have disappeared, leaving only their concepts behind them.

I find this intensional language interesting, for it illustrates what it would be like to be able to give the modalities free rein. But this repudiation of the concrete and extensional is a more radical move, in general, than a mere comparison of 43-3 with 43-2 might suggest. The strangeness of the intensional language becomes more evident when we try to reformulate statements such as these:

(1) The number of planets is a power of three,
(2) The wives of two of the directors are deaf.

In the familiar logic, (1) and (2) would be analyzed in part as follows:

(3) $(\exists n)$ (n is a natural number . the number of planets $= 3^n$),
(4) $(\exists x)(\exists y)(\exists z)(\exists w)$[$x$ is a director . y is a director . $\sim (x = y)$. z is wife of x . w is wife of y . z is deaf . w is deaf].

But the formulation (3) depends on there being numbers (extensions, presumably classes of classes) as values of the bound variable; and the formulation (4) depends on there being persons (extensions, individuals) as values of the four bound variables. Failing such values, (3) and (4) would have to be reformulated in terms of number concepts and individual concepts. The logical predicate '=' of identity in (3) and (4) would thereupon have to give way to a logical predicate of extensional equivalence of concepts. The logical predicate 'is a natural number' in (3) would have to give way to a logical predicate having the sense 'is a natural-number-concept'. The empirical predicates 'is a director', 'is wife of', and 'is deaf', in (4), would have to give way to some new predicates whose senses are more readily imagined than put into words. These examples do not prove your language-structure inadequate, but they give some hint of the unusual character which a development of it adequate to general purposes would have to assume.

The first important point to be noticed in Quine's statement is that he agrees that the form of modal language explained in the present chapter "is indeed an effective way of reconciling quantification and modality". Some readers of Quine's article believed that it proved the impossibility of a logical system combining modalities with variables. Quine's statement now shows that this is not the case.

However, there are still some serious problems involved. Quine, while admitting the possibility of modal systems with quantification, believes that these systems have certain peculiar features which he regards as disadvantages. Let us now examine these problems.

I have previously explained (at the beginning of § 10) that I agree with Quine's view that an author who uses variables of some kind thereby indicates that he recognizes those entities which are values of the variables. (I have simultaneously expressed some doubts concerning the advisability of applying the term 'ontology' to this recognition; but for our present discussion we may leave aside this question.) It is the counterpart of this thesis that is of importance for our problem; it says that, if someone uses a language which does not contain any variables with certain entities as values, he thereby indicates that he does not recognize these entities or at least that he does not intend to speak about them as long as he restricts himself to the use of this language. In a certain sense, I can agree also with this thesis. As an example, let us compare the following two languages S_P and S_P'. Let S_P be the ordinary language of physics (§ 19). It contains variables which have real numbers, both rational and irrational, as values. Suppose somebody proposes another language S_P' for physics which contains variables for rational numbers, but no variables to whose values irrational numbers belong. Here I would be willing to say, like Quine, that the user of this language S_P' excludes or "repudiates" the irrational numbers and that these numbers "have disappeared" from the universe of discourse. Now Quine says that the variables in the modal language have as values only intensions, not extensions, and that therefore, as far as this language is concerned, all extensions, for example, classes and "the individuals of the concrete world", "have disappeared". With this I cannot agree. At the first glance, the situation here may seem to be similar to that in the example of the irrational numbers; but actually it is fundamentally different.

In order to clarify the situation, we shall contrast in the following discussion our two language systems, the extensional language S_1 and the modal language S_2. We shall further consider the following two extended languages. The language S_1' is extensional like S_1 but contains additional kinds of variables, say '*f*', '*g*', etc., for which predicators of level one (and degree one) are substitutable, '*m*', '*n*', etc., for predicators of level two, and '*p*', '*q*', etc., for sentences. The language S_2' is constructed from S_1' by the addition of 'N'; hence it is a modal language like S_2. According to Quine's view, the values of '*f*' in S_2' are not classes but properties, because '$(f)(f \equiv f)$' holds. In the extensional system S_1', on the other hand, we have only '$(f)(f \equiv f)$'. Therefore, Quine will presumably regard classes as the values of '*f*' in this system, as he does for the variables of his extensional system ML (see above, § 25). Similarly, Quine says that the values of individual variables (e.g., 'x') in modal systems like S_2 and S_2' are individual concepts; on the other hand, he presumably regards individuals (concrete things or positions) as the values of individual variables in extensional systems like S_1 and S_1'. Now the decisive point is the following: As explained previously (§ 35), there is no objection against regarding designators in a modal language as names of intensions and regarding variables as having intensions as values, provided we are not misled by this formulation into the erroneous conception that the extensions have disappeared from the universe of discourse of the language. As explained earlier (§ 27), it is not possible for a predicator in an interpreted language to possess only an extension and not an intension or, in customary terms, to refer only to a class and not to a property. Similarly, it is impossible for a variable to be merely a class variable and not also a property variable. On the other hand, it is, of course, possible for a variable to have as values only properties and no relations, or only rational numbers and no irrational numbers. This shows the difference between the two cases. For example, the so-called class variables in the system PM′ (e.g., 'α') are, as we have seen (§ 27), also property variables, that is to say, they have properties as value intensions. The same holds now for variables like 'f' in S_1'. Languages of Quine's form ML′ or of Russell's form PM′ or of our form S_1' speak also about properties. The restriction of these extensional languages in comparison with modal languages like S_2' consists merely in the fact that whatever is said in any of these languages about a property is either true for all equivalent properties or false for all equivalent properties; in technical terms, all properties of properties expressible in these languages (by a matrix with a free variable of the kind mentioned) are extensional. This makes it possible to paraphrase all sentences of these languages in terms of classes. An analogous result holds for individual variables. These variables in an extensional language like S_1 and S_1' refer not only to individuals but also, and even primarily, to individual concepts. The restriction is again merely this: Whatever is said in these languages about individual concepts is either true for all equivalent individual concepts or false for all of them; in technical terms, it is extensional. Therefore, whatever is said in these languages about individual concepts can be paraphrased in terms of individuals.

Although the sentences of an extensional language (S_1 or S_1') can thus be interpreted as speaking about individuals and classes, they can be translated into the corresponding modal language (S_2 or S_2', respectively). This translation fulfils not only the requirement of L-equivalence but also the requirement of intensional isomorphism, the strictest requirement that any translation can fulfil (§ 14). Any given sentence in S_1' is translated into S_2' by that sentence itself, that is, by the same sequence of signs now taken as signs in S_2'. Any two corresponding designators, that is, any designator in S_1' and the same expression in S_2', are L-equivalent to one another. This follows from the following two results:

(i) The rules of designation for the descriptive signs are the same in both systems S_1' and S_2' (for example, the rules 1-2 for primitive predicators).

(ii) Any sentence in S_1' has the same range in both systems S_1' and S_2' (see 41-4 concerning S_1 and S_2). Since the range is the same, the truth-conditions are the same; therefore, the sentence means exactly the same in S_2' as in S_1'.

Thus the decisive difference between the situation here and that in the earlier example concerning the irrational numbers becomes clear. In the transition from S_P to S_P' the irrational numbers actually disappear, because a sentence in S_P of the form 'there is an irrational number such that . . .' is not translatable into S_P'. On the other hand, in the transition from an extensional to a modal language the individuals and classes do by no means disappear. A sentence in S_1 (or S_1') which says that there is an individual of a certain kind is translatable into S_2 (or S_2'); and a sentence in S_1' which says that there is a class of a certain kind is translatable into S_2'.

In order to illustrate this result by an example, let us take Quine's sentence (2). Since this sentence requires only individual variables, it can be translated into S_1. Let us assume that S_1 contains the following predicators, either as primitive signs or as defined in a suitable way: 'W' for the relation Wife, 'D' for the property Director, and 'F' for the property Deaf. Then (2) is translated into S_1 by the following sentence:

(5) '$(\exists x)(\exists y)(\exists z)(\exists w)[Dx \bullet Dy \bullet \sim(x \equiv y) \bullet Wzx \bullet Wwy \bullet Fz \bullet Fw]$'.

Now this same sentence is also the translation of (2) into S_2. It would be an error to think that it was necessary for the translation into S_2 either to use new predicators or to assign a new meaning to the old predicators, as though, for example, 'Dx' in S_1 said that the individual x has the property Director while 'Dx' in S_2 said that the individual concept x has a strange new property somehow analogous but not quite the same as the property Director. The matrix 'Dx' expresses in both languages the property Director; it may be defined in both languages in exactly the same way. Suppose a speaker X_1 uses the language S_1 and X_2 uses S_2. Then the question of whether a given full sentence, say 'Db', is true, may be decided by both speakers in the same way. Both confirm or disconfirm this sentence on the basis of observations of the person b, using the same empirical criteria for the property Director. Nothing in the semantical analysis of this sentence or in the procedure of empirical confirmation or in the expectation of possible future experiences implied by the sentence needs to be different for the two speakers. The same holds for the existential sentence (5) and for any other sentence occurring in both languages. Therefore, I cannot agree with the view that, while the speaker X_1 recognizes the individuals of the concrete world, they have disappeared for X_2, leaving only their concepts behind them.

The situation with respect to Quine's other example (1) is analogous, except that cardinal numbers are involved and therefore a variable of second level, say '*n*', is used. We have seen earlier (§ 27) that, for the introduction of particular cardinal numbers and of the general concept of cardinal number, it is not necessary to use special class expressions and class variables, as Frege and Russell did; we may, instead, regard cardinal numbers as properties of second level or, rather, introduce cardinal number expressions as predicators of second level, whose intensions are properties of second level and whose extensions are classes of second level. Equality of cardinal numbers is then expressed with the help of '$\equiv$'. Thus we translated the sentence

(6) 'the number of planets = 9'

into the following sentence of S_1':

(7) '$Nc{}^{\backprime}P \equiv 9$'.

Similarly, Quine's sentence (1) can be translated into S_1' as follows, if we assume that exponentiation has been defined by a suitable procedure (analogous to that of Cantor or Russell, [P.M.], Vol. II, *116):

(8) '$(\exists n)[NC(n) \bullet Nc{}^{\backprime}P \equiv 3^n]$'.

(If we wish to say that *n* is finite, we may use the concept of inductive cardinal number with a definition analogous to Russell's). Here, again, the given sentence (1) can likewise be translated into the modal language S_2', namely, by the same sentence (8), hence without the use of any strange new concepts. The translation is by no means dependent upon the occurrence of class variables as distinct from property variables. '$NC(n)$' means in S_2', just as in S_1', that *n* is a cardinal number; thus in S_2', just as in S_1', sentences like '$NC(2)$' and '$NC(Nc{}^{\backprime}P)$' are L-true. That the sentence (8) has in S_2' the same factual content as in S_1' is seen by considerations similar to those concerning the previous example (5). The same astronomical observations confirm the sentence in the one as in the other language; it gives rise to the same expectations of future observations in both languages. Thus there cannot be any difference in meaning.

The preceding discussion shows that a modal language is not inadequate in comparison with the corresponding extensional language, that is to say that we can express in the former whatever is expressed in the latter. (So much Quine seems to admit.) We have seen, moreover, that the expressions used in a modal language for translations from the extensional language do not have any unusual character with respect to either their form or their meaning. Every designator and every sentence in the extensional language has exactly the same meaning in the modal language—more exactly speaking, it has both the same intension and the same extension. The world of concrete things and the conceptual world of numbers are dealt with in the modal language just as well as in the extensional one. In order to see correctly the functions of these languages, and generally of any languages, it is essential to abandon the old prejudice that a predicator must stand either for a class or for a property but cannot stand for both and that an individual expression must stand either for an individual or for an individual concept but cannot stand for both. To understand how language works, we must realize that every designator has both an intension and an extension.

§ 45. Conclusions

The main conclusions of the discussions in this book are briefly summarized. The difference between the two operations—understanding the meaning of a given expression and investigating whether and how it applies to the actual state of the world—suggests a distinction between two different semantical factors, which our method tries to explicate by the concepts of the intension and the extension of an expression.

The chief purpose of this book is to develop a method for the analysis of meaning in language, hence a semantical method. We may distinguish two operations with respect to a given linguistic expression, in particular, a (declarative) sentence and its parts. The first operation is the analysis of the expression with the aim of understanding it, of grasping its meaning. This operation is a logical or semantical one; in its technical form it is based on the semantical rules concerning the given expression. The second operation consists in investigations concerning the factual situation referred to by the given expression. Its aim is the establishment of factual truth. This operation is not of a purely logical, but of an empirical, nature. We can distinguish two sides or factors in the given expression with regard to these two operations. The first factor is that side of the expression which we can establish by the first operation alone, that is, by understanding without using factual knowledge. This is what is usually called the meaning of the expression. In our method it is explicated by the technical concept of intension. The second factor is established by both operations together. Knowing the meaning, we discover by an investigation of facts to which locations, if any, the expression applies in the actual state of the world. This factor is explicated in our method by the technical concept of extension. Thus, for every expression which we can understand, there is the question of meaning and the question of actual application; therefore, the expression has primarily an intension and secondarily an extension.

The method of intension and extension stands in contrast to the customary method of the name-relation. The basic weakness of the latter method is its failure to realize the fundamental distinction between meaning and application. This leads to the conception that an expression must be the name of exactly one of the two semantical factors involved. For example, properties and classes are regarded as entities of equal standing; this leads to the view that a language ought to contain both names of properties and names of classes. This conception is the ultimate source of the various difficulties which we found involved in the method of the name-relation. They center around the well-known difficulty which we have called the antinomy of the name-relation. We have seen how the various methods of keeping the name-relation but avoiding the antinomy lead either to great complications in the language structure or to serious restrictions in the use of the language or in the application of the semantical method.

The formulations in terms of 'extension' and 'intension', 'class' and 'property', etc., seem to refer to two kinds of entities in each type. We have seen, however, that, in fact, no such duplication of entities is presupposed by our method and that those formulations involve only a convenient duplication of modes of speech. As it was shown to be unnecessary to use different expressions for classes and properties in a symbolic object language, it likewise turned out to be unnecessary to use those pairs of terms in the word language as a metalanguage. A new metalanguage was constructed, in which instead of the pair of phrases 'the class Human' and 'the property Human' only the neutral term 'Human' is used. It was shown that the ordinary formulations can be translated into this neutral metalanguage and that the latter language preserves all previous distinctions, though in different formulations.

Our semantical method also helps in the clarification of the problems of the modalities. It suggests a certain interpretation of the logical modalities which supplies a suitable basis for a system of modal logic. In particular, the distinction between intensions and extensions enables us to overcome the difficulties involved in combining modalities with quantified variables.

The different conceptions of other authors discussed in this book, for instance, those of Frege, Russell, Church, and Quine, concerning semantical problems, that is, problems of meaning, extension, naming, denotation, and the like, have sometimes been regarded as different theories so that one of them at most could be right while all others must be false. I regard these conceptions and my own rather as different methods, methods of semantical analysis characterized chiefly by the concepts used. Of course, once a method has been chosen, the question of whether or not certain results are valid on its basis is a theoretical one. But there is hardly any question of this kind on which I disagree with one of the other authors. Our differences are mainly practical differences concerning the choice of a method for semantical analysis. Methods, unlike logical statements, are never final. For any method of semantical analysis which someone proposes, somebody else will find improvements, that is, changes which will seem preferable to him and many others. This will certainly hold for the method which I have proposed here, no less than for the others.

Let me conclude our discussions by borrowing the words with which Russell concludes his paper.[10] It seems to me that his remarks, although written more than forty years ago, still apply to the present situation (except, perhaps, that instead of 'the true theory' I might prefer to say 'the best method'):

"Of the many other consequences of the view I have been advocating, I will say nothing. I will only beg the reader not to make up his mind against the view—as he might be tempted to do, on account of its apparently excessive complication—until he has attempted to construct a theory of his own on the subject of denotation. This attempt, I believe, will convince him that, whatever the true theory may be, it cannot have such a simplicity as one might have expected beforehand."

[10] [Denoting], p. 493.

SUPPLEMENT

This Supplement consists of five previously published articles. How they are related to the main body of the book is indicated in my Preface to the Second Edition. For the original places of their publication, see the starred items in the Bibliography.

A. EMPIRICISM, SEMANTICS, AND ONTOLOGY*

1. The Problem of Abstract Entities

Empiricists are in general rather suspicious with respect to any kind of abstract entities like properties, classes, relations, numbers, propositions, etc. They usually feel much more in sympathy with nominalists than with realists (in the medieval sense). As far as possible they try to avoid any reference to abstract entities and to restrict themselves to what is sometimes called a nominalistic language, i.e., one not containing such references. However, within certain scientific contexts it seems hardly possible to avoid them. In the case of mathematics, some empiricists try to find a way out by treating the whole of mathematics as a mere calculus, a formal system for which no interpretation is given or can be given. Accordingly, the mathematician is said to speak not about numbers, functions, and infinite classes, but merely about meaningless symbols and formulas manipulated according to given formal rules. In physics it is more difficult to shun the suspected entities, because the language of physics serves for the communication of reports and predictions and hence cannot be taken as a mere calculus. A physicist who is suspicious of abstract entities may perhaps try to declare a certain part of the language of physics as uninterpreted and uninterpretable, that part which refers to real numbers as space-time coordinates or as values of physical magnitudes, to functions, limits, etc. More probably he will just speak about all these things like anybody else but with an uneasy conscience, like a man who in his everyday life does with qualms many things which are not in accord with the high moral principles he professes on Sundays. Recently the problem of abstract entities has arisen again in connection with semantics, the theory

* I have made here some minor changes in the formulations to the effect that the term "framework" is now used only for the system of linguistic expressions, and not for the system of the entities in question.

of meaning and truth. Some semanticists say that certain expressions designate certain entities, and among these designated entities they include not only concrete material things but also abstract entities, e.g., properties as designated by predicates and propositions as designated by sentences.[1] Others object strongly to this procedure as violating the basic principles of empiricism and leading back to a metaphysical ontology of the Platonic kind.

It is the purpose of this article to clarify this controversial issue. The nature and implications of the acceptance of a language referring to abstract entities will first be discussed in general; it will be shown that using such a language does not imply embracing a Platonic ontology but is perfectly compatible with empiricism and strictly scientific thinking. Then the special question of the role of abstract entities in semantics will be discussed. It is hoped that the clarification of the issue will be useful to those who would like to accept abstract entities in their work in mathematics, physics, semantics, or any other field; it may help them to overcome nominalistic scruples.

2. Linguistic Frameworks

Are there properties, classes, numbers, propositions? In order to understand more clearly the nature of these and related problems, it is above all necessary to recognize a fundamental distinction between two kinds of questions concerning the existence or reality of entities. If someone wishes to speak in his language about a new kind of entities, he has to introduce a system of new ways of speaking, subject to new rules; we shall call this procedure the construction of a linguistic *framework* for the new entities in question. And now we must distinguish two kinds of questions of existence: first, questions of the existence of certain entities of the new kind *within the framework;* we call them *internal questions;* and second, questions concerning the existence or reality *of the system of entities as a whole*, called *external questions*. Internal questions and possible answers to them are formulated with the help of the new forms of expressions. The answers may be found either by purely logical methods or by empirical methods, depending upon whether the framework is a logical or a factual one. An external question is of a problematic character which is in need of closer examination.

The world of things. Let us consider as an example the simplest kind of entities dealt with in the everyday language: the spatio-temporally or-

[1] The terms "sentence" and "statement" are here used synonymously for declarative (indicative, propositional) sentences.

dered system of observable things and events. Once we have accepted the thing language with its framework for things, we can raise and answer internal questions, e.g., "Is there a white piece of paper on my desk?", "Did King Arthur actually live?", "Are unicorns and centaurs real or merely imaginary?", and the like. These questions are to be answered by empirical investigations. Results of observations are evaluated according to certain rules as confirming or disconfirming evidence for possible answers. (This evaluation is usually carried out, of course, as a matter of habit rather than a deliberate, rational procedure. But it is possible, in a rational reconstruction, to lay down explicit rules for the evaluation. This is one of the main tasks of a pure, as distinguished from a psychological, epistemology.) The concept of reality occurring in these internal questions is an empirical, scientific, non-metaphysical concept. To recognize something as a real thing or event means to succeed in incorporating it into the system of things at a particular space-time position so that it fits together with the other things recognized as real, according to the rules of the framework.

From these questions we must distinguish the external question of the reality of the thing world itself. In contrast to the former questions, this question is raised neither by the man in the street nor by scientists, but only by philosophers. Realists give an affirmative answer, subjective idealists a negative one, and the controversy goes on for centuries without ever being solved. And it cannot be solved because it is framed in a wrong way. To be real in the scientific sense means to be an element of the system; hence this concept cannot be meaningfully applied to the system itself. Those who raise the question of the reality of the thing world itself have perhaps in mind not a theoretical question as their formulation seems to suggest, but rather a practical question, a matter of a practical decision concerning the structure of our language. We have to make the choice whether or not to accept and use the forms of expression in the framework in question.

In the case of this particular example, there is usually no deliberate choice because we all have accepted the thing language early in our lives as a matter of course. Nevertheless, we may regard it as a matter of decision in this sense: we are free to choose to continue using the thing language or not; in the latter case we could restrict ourselves to a language of sense-data and other "phenomenal" entities, or construct an alternative to the customary thing language with another structure, or, finally, we could refrain from speaking. If someone decides to accept the thing language, there is no objection against saying that he has accepted the world of things. But this must not be interpreted as if it meant his acceptance of a *belief* in the reality of the thing world; there is no such belief or assertion or assumption, because it is not a theoretical question. To accept the thing world means nothing more than to accept a certain form of language, in other words, to accept rules for forming statements and for testing, accepting, or rejecting them. The acceptance of the thing language leads, on the basis of observations made, also to the acceptance, belief, and assertion of certain statements. But the thesis of the reality of the thing world cannot be among these statements, because it cannot be formulated in the thing language or, it seems, in any other theoretical language.

The decision of accepting the thing language, although itself not of a cognitive nature, will nevertheless usually be influenced by theoretical knowledge, just like any other deliberate decision concerning the acceptance of linguistic or other rules. The purposes for which the language is intended to be used, for instance, the purpose of communicating factual knowledge, will determine which factors are relevant for the decision. The efficiency, fruitfulness, and simplicity of the use of the thing language may be among the decisive factors. And the questions concerning these qualities are indeed of a theoretical nature. But these questions cannot be identified with the question of realism. They are not yes-no questions but questions of degree. The thing language in the customary form works indeed with a high degree of efficiency for most purposes of everyday life. This is a matter of fact, based upon the content of our experiences. However, it would be wrong to describe this situation by saying: "The fact of the efficiency of the thing language is confirming evidence for the reality of the thing world"; we should rather say instead: "This fact makes it advisable to accept the thing language".

The system of numbers. As an example of a system which is of a logical rather than a factual nature let us take the system of natural numbers. The framework for this system is constructed by introducing into the language new expressions with suitable rules: (1) numerals like "five" and sentence forms like "there are five books on the table"; (2) the general term "number" for the new entities, and sentence forms like "five is a number"; (3) expressions for properties of numbers (e.g., "odd", "prime"), relations (e.g., "greater than"), and functions (e.g., "plus"), and sentence forms like "two plus three is five"; (4) numerical variables ("*m*", "*n*", etc.) and quantifiers for universal sentences ("for every *n*, . . .") and existential sentences ("there is an *n* such that . . .") with the customary deductive rules.

Here again there are internal questions, e.g., "Is there a prime number greater than a hundred?" Here, however, the answers are found, not by empirical investigation based on observations, but by logical analysis based on the rules for the new expressions. Therefore the answers are here analytic, i.e., logically true.

What is now the nature of the philosophical question concerning the existence or reality of numbers? To begin with, there is the internal question which, together with the affirmative answer, can be formulated in the new terms, say, by "There are numbers" or, more explicitly, "There is an *n* such that *n* is a number". This statement follows from the analytic statement "five is a number" and is therefore itself analytic. Moreover, it is rather trivial (in contradistinction to a statement like "There is a prime number greater than a million", which is likewise analytic but far from trivial), because it does not say more than that the new system is not empty; but this is immediately seen from the rule which states that words like "five" are substitutable for the new variables. Therefore nobody who meant the question "Are there numbers?" in the internal sense would either assert or even seriously consider a negative answer. This makes it plausible to assume that those philosophers who treat the question of the existence of numbers as a serious philosophical problem and offer lengthy arguments on either side, do not have in mind the internal question. And, indeed, if we were to ask them: "Do you mean the question as to whether the framework of numbers, *if* we were to accept it, would be found to be empty or not?", they would probably reply: "Not at all; we mean a question *prior* to the acceptance of the new framework". They might try to explain what they mean by saying that it is a question of the ontological status of numbers; the question whether or not numbers have a certain metaphysical characteristic called reality (but a kind of ideal reality, different from the material reality of the thing world) or subsistence or status of "independent entities". Unfortunately, these philosophers have so far not given a formulation of their question in terms of the common scientific language. Therefore our judgment must be that they have not succeeded in giving to the external question and to the possible answers any cognitive content. Unless and until they supply a clear cognitive interpretation, we are justified in our suspicion that their question is a pseudo-question, that is, one disguised in the form of a theoretical question while in fact it is non-theoretical; in the present case it is the practical problem whether or not to incorporate into the language the new linguistic forms which constitute the framework of numbers.

The system of propositions. New variables, "*p*", "*q*", etc., are introduced with a rule to the effect that any (declarative) sentence may be substituted for a variable of this kind; this includes, in addition to the sentences of the original thing language, also all general sentences with variables of any kind which may have been introduced into the language. Further, the general term "proposition" is introduced. "*p* is a proposition" may be defined by "*p* or not *p*" (or by any other sentence form yielding only analytic sentences). Therefore, every sentence of the form ". . . is a proposition" (where any sentence may stand in the place of the dots) is analytic. This holds, for example, for the sentence:

(*a*) "Chicago is large is a proposition".

(We disregard here the fact that the rules of English grammar require not a sentence but a that-clause as the subject of another sentence; accordingly, instead of (*a*) we should have to say "That Chicago is large is a proposition".) Predicates may be admitted whose argument expressions are sentences; these predicates may be either extensional (e.g., the customary truth-functional connectives) or not (e.g., modal predicates like "possible", "necessary", etc.). With the help of the new variables, general sentences may be formed, e.g.,

(*b*) "For every *p*, either *p* or not-*p*".

(*c*) "There is a *p* such that *p* is not necessary and not-*p* is not necessary".

(*d*) "There is a *p* such that *p* is a proposition".

(*c*) and (*d*) are internal assertions of existence. The statement "There are propositions" may be meant in the sense of (*d*); in this case it is analytic (since it follows from (*a*)) and even trivial. If, however, the statement is meant in an external sense, then it is non-cognitive.

It is important to notice that the system of rules for the linguistic expressions of the propositional framework (of which only a few rules have here been briefly indicated) is sufficient for the introduction of the framework. Any further explanations as to the nature of the propositions (i.e., the elements of the system indicated, the values of the variables "*p*", "*q*", etc.) are theoretically unnecessary because, if correct, they follow from the rules. For example, are propositions mental events (as in Russell's theory)? A look at the rules shows us that they are not, because otherwise existential statements would be of the form: "If the mental state of the person in question fulfils such and such conditions, then there is a *p* such that . . .". The fact that no references to mental conditions occur in existential statements (like (*c*), (*d*), etc.) shows that propositions are not mental entities. Further, a statement of the existence of linguistic entities (e.g., expressions, classes of expressions, etc.) must contain a reference to a language. The fact that no such reference occurs in the existential statements here, shows that propositions are not linguistic entities. The fact that in these statements no reference to a subject (an observer or knower) occurs (nothing like: "There is a *p* which is necessary for Mr. *X*"), shows that the propositions (and their properties, like necessity, etc.) are not subjective. Although characterizations of these or similar kinds are, strictly speaking, unnecessary, they may nevertheless be practically useful. If they are given, they should be understood, not as ingredient parts of the system, but merely as marginal notes with the purpose of supplying to the reader helpful hints or convenient pictorial associations which may make his learning of the use of the expressions easier than the bare system of the rules would do. Such a characterization is analogous to an extra-systematic explanation which a physicist sometimes gives to the beginner. He might, for example, tell him to imagine the atoms of a gas as small balls rushing around with great speed, or the electromagnetic field and its oscillations as quasi-elastic tensions and vibrations in an ether. In fact, however, all that can accurately be said about atoms or the field is implicitly contained in the physical laws of the theories in question.[2]

The system of thing properties. The thing language contains words like "red", "hard", "stone", "house", etc., which are used for describing what things are like. Now we may introduce new variables, say "*f*", "*g*", etc., for which those words are substitutable and furthermore the general term "property". New rules are laid down which admit sentences like "Red is a property", "Red is a color", "These two pieces of paper have at least one

[2] In my book *Meaning and Necessity* (Chicago, 1947) I have developed a semantical method which takes propositions as entities designated by sentences (more specifically, as intensions of sentences). In order to facilitate the understanding of the systematic development, I added some informal, extra-systematic explanations concerning the nature of propositions. I said that the term "proposition" "is used neither for a linguistic expression nor for a subjective, mental occurrence, but rather for something objective that may or may not be exemplified in nature. . . . We apply the term 'proposition' to any entities of a certain logical type, namely, those that may be expressed by (declarative) sentences in a language" (p. 27). After some more detailed discussions concerning the relation between propositions and facts, and the nature of false propositions, I added: "It has been the purpose of the preceding remarks to facilitate the understanding of our conception of propositions. If, however, a reader should find these explanations more puzzling than clarifying, or even unacceptable, he may disregard them" (p. 31) (that is, disregard these extra-systematic explanations, not the whole theory of the propositions as intensions of sentences, as one reviewer understood). In spite of this warning, it seems that some of those readers who were puzzled by the explanations, did not disregard them but thought that by raising objections against them they could refute the theory. This is analogous to the procedure of some laymen who by (correctly) criticizing the ether picture or other visualizations of physical theories, thought they had refuted those theories. Perhaps the discussions in the present paper will help in clarifying the role of the system of linguistic rules for the introduction of a framework for entities on the one hand, and that of extra-systematic explanations concerning the nature of the entities on the other.

color in common" (i.e., "There is an *f* such that *f* is a color, and . . ."). The last sentence is an internal assertion. It is of an empirical, factual nature. However, the external statement, the philosophical statement of the reality of properties—a special case of the thesis of the reality of universals—is devoid of cognitive content.

The systems of integers and rational numbers. Into a language containing the framework of natural numbers we may introduce first the (positive and negative) integers as relations among natural numbers and then the rational numbers as relations among integers. This involves introducing new types of variables, expressions substitutable for them, and the general terms "integer" and "rational number".

The system of real numbers. On the basis of the rational numbers, the real numbers may be introduced as classes of a special kind (segments) of rational numbers (according to the method developed by Dedekind and Frege). Here again a new type of variables is introduced, expressions substitutable for them (e.g., "$\sqrt{2}$"), and the general term "real number".

The spatio-temporal coordinate system for physics. The new entities are the space-time points. Each is an ordered quadruple of four real numbers, called its coordinates, consisting of three spatial and one temporal coordinates. The physical state of a spatio-temporal point or region is described either with the help of qualitative predicates (e.g., "hot") or by ascribing numbers as values of a physical magnitude (e.g., mass, temperature, and the like). The step from the system of things (which does not contain space-time points but only extended objects with spatial and temporal relations between them) to the physical coordinate system is again a matter of decision. Our choice of certain features, although itself not theoretical, is suggested by theoretical knowledge, either logical or factual. For example, the choice of real numbers rather than rational numbers or integers as coordinates is not much influenced by the facts of experience but mainly due to considerations of mathematical simplicity. The restriction to rational coordinates would not be in conflict with any experimental knowledge we have, because the result of any measurement is a rational number. However, it would prevent the use of ordinary geometry (which says, e.g., that the diagonal of a square with the side 1 has the irrational value $\sqrt{2}$) and thus lead to great complications. On the other hand, the decision to use three rather than two or four spatial coordinates is strongly suggested, but still not forced upon us, by the result of common observations. If certain events allegedly observed in spiritualistic séances, e.g., a ball moving out of a sealed box, were confirmed beyond any reasonable doubt, it might seem advisable to use four spatial coordinates. Internal questions are here, in general, empirical questions to be answered by empirical investigations. On the other hand, the external questions of the reality of physical space and physical time are pseudo-questions. A question like "Are there (really) space-time points?" is ambiguous. It may be meant as an internal question; then the affirmative answer is, of course, analytic and trivial. Or it may be meant in the external sense: "Shall we introduce such and such forms into our language?"; in this case it is not a theoretical but a practical question, a matter of decision rather than assertion, and hence the proposed formulation would be misleading. Or finally, it may be meant in the following sense: "Are our experiences such that the use of the linguistic forms in question will be expedient and fruitful?" This is a theoretical question of a factual, empirical nature. But it concerns a matter of degree; therefore a formulation in the form "real or not?" would be inadequate.

3. What Does Acceptance of a Kind of Entities Mean?

Let us now summarize the essential characteristics of situations involving the introduction of a new kind of entities, characteristics which are common to the various examples outlined above.

The acceptance of a new kind of entities is represented in the language by the introduction of a framework of new forms of expressions to be used according to a new set of rules. There may be new names for particular entities of the kind in question; but some such names may already occur in the language before the introduction of the new framework. (Thus, for example, the thing language contains certainly words of the type of "blue" and "house" before the framework of properties is introduced; and it may contain words like "ten" in sentences of the form "I have ten fingers" before the framework of numbers is introduced.) The latter fact shows that the occurrence of constants of the type in question—regarded as names of entities of the new kind after the new framework is introduced—is not a sure sign of the acceptance of the new kind of entities. Therefore the introduction of such constants is not to be regarded as an essential step in the introduction of the framework. The two essential steps are rather the following. First, the introduction of a general term, a predicate of higher level, for the new kind of entities, permitting us to say of any particular entity that it belongs to this kind (e.g., "Red is a *property*", "Five is a *number*"). Second, the introduction of variables of the new type. The new entities are values of these variables; the constants (and the closed compound expressions, if any) are substitutable for the variables.[3] With the help of the variables, general sentences concerning the new entities can be formulated.

After the new forms are introduced into the language, it is possible to formulate with their help internal questions and possible answers to them. A question of this kind may be either empirical or logical; accordingly a true answer is either factually true or analytic.

From the internal questions we must clearly distinguish external questions, i.e., philosophical questions concerning the existence or reality of the total system of the new entities. Many philosophers regard a question of this kind as an ontological question which must be raised and answered *before* the introduction of the new language forms. The latter introduction, they believe, is legitimate only if it can be justified by an ontological insight supplying an affirmative answer to the question of reality. In contrast to this view, we take the position that the introduction of the new ways of speaking does not need any theoretical justification because it does not imply any assertion of reality. We may still speak (and have done so) of "the acceptance of the new entities" since this form of speech is customary; but one must keep in mind that this phrase does not mean for us anything more than acceptance of the new framework, i.e., of the new linguistic forms. Above all, it must not be interpreted as referring to an assumption, belief, or assertion of "the reality of the entities". There is no such assertion. An alleged statement of the reality of the system of entities is a pseudo-statement without cognitive content. To be sure, we have to face at this point an important question; but it is a practical, not a theoretical question; it is the question of whether or not to accept the new linguistic forms. The acceptance cannot be judged as being either true or false because it is not an assertion. It can only be judged as being more or less expedient, fruitful, conducive to the aim for which the language is intended. Judgments of this kind supply the motivation for the decision of accepting or rejecting the kind of entities.[4]

Thus it is clear that the acceptance of a linguistic framework must not be regarded as implying a metaphysical doctrine concerning the reality of the entities in question. It seems to me due to a neglect of this important

[3] W. V. Quine was the first to recognize the importance of the introduction of variables as indicating the acceptance of entities. "The ontology to which one's use of language commits him comprises simply the objects that he treats as falling . . . within the range of values of his variables" ([Notes], p. 118; compare also his [Designation] and [Universals]).

[4] For a closely related point of view on these questions see the detailed discussions in Herbert Feigl, "Existential Hypotheses", *Philosophy of Science*, 17 (1950), 35–62.

distinction that some contemporary nominalists label the admission of variables of abstract types as "Platonism".[5] This is, to say the least, an extremely misleading terminology. It leads to the absurd consequence, that the position of everybody who accepts the language of physics with its real number variables (as a language of communication, not merely as a calculus) would be called Platonistic, even if he is a strict empiricist who rejects Platonic metaphysics.

A brief historical remark may here be inserted. The non-cognitive character of the questions which we have called here external questions was recognized and emphasized already by the Vienna Circle under the leadership of Moritz Schlick, the group from which the movement of logical empiricism originated. Influenced by ideas of Ludwig Wittgenstein, the Circle rejected both the thesis of the reality of the external world and the thesis of its irreality as pseudo-statements;[6] the same was the case for both the thesis of the reality of universals (abstract entities, in our present terminology) and the nominalistic thesis that they are not real and that their alleged names are not names of anything but merely *flatus vocis*. (It is obvious that the apparent negation of a pseudo-statement must also be a pseudo-statement.) It is therefore not correct to classify the members of the Vienna Circle as nominalists, as is sometimes done. However, if we look at the basic anti-metaphysical and pro-scientific attitude of most nominalists (and the same holds for many materialists and realists in the modern sense), disregarding their occasional pseudo-theoretical formulations, then it is, of course, true to say that the Vienna Circle was much closer to those philosophers than to their opponents.

[5] Paul Bernays, "Sur le platonisme dans les mathématiques" (*L'Enseignement math.*, 34 (1935), 52–69). W. V. Quine, see previous footnote and a recent paper [What]. Quine does not acknowledge the distinction which I emphasize above, because according to his general conception there are no sharp boundary lines between logical and factual truth, between questions of meaning and questions of fact, between the acceptance of a language structure and the acceptance of an assertion formulated in the language. This conception, which seems to deviate considerably from customary ways of thinking, will be explained in his article [Semantics]. When Quine in the article [What] classifies my logicistic conception of mathematics (derived from Frege and Russell) as "platonic realism" (p. 33), this is meant (according to a personal communication from him) not as ascribing to me agreement with Plato's metaphysical doctrine of universals, but merely as referring to the fact that I accept a language of mathematics containing variables of higher levels. With respect to the basic attitude to take in choosing a language form (an "ontology" in Quine's terminology, which seems to me misleading), there appears now to be agreement between us: "the obvious counsel is tolerance and an experimental spirit" ([What], p. 38).

[6] See Carnap, *Scheinprobleme in der Philosophie; das Fremdpsychische und der Realismusstreit*, Berlin, 1928. Moritz Schlick, *Positivismus und Realismus*, reprinted in *Gesammelte Aufsätze*, Wien, 1938.

4. Abstract Entities in Semantics

The problem of the legitimacy and the status of abstract entities has recently again led to controversial discussions in connection with semantics. In a semantical meaning analysis certain expressions in a language are often said to designate (or name or denote or signify or refer to) certain extra-linguistic entities.[7] As long as physical things or events (e.g., Chicago or Caesar's death) are taken as designata (entities designated), no serious doubts arise. But strong objections have been raised, especially by some empiricists, against abstract entities as designata, e.g., against semantical statements of the following kind:

(1) "The word 'red' designates a property of things";

(2) "The word 'color' designates a property of properties of things";

(3) "The word 'five' designates a number";

(4) "The word 'odd' designates a property of numbers";

(5) "The sentence 'Chicago is large' designates a proposition".

Those who criticize these statements do not, of course, reject the use of the expressions in question, like "red" or "five"; nor would they deny that these expressions are meaningful. But to be meaningful, they would say, is not the same as having a meaning in the sense of an entity designated. They reject the belief, which they regard as implicitly presupposed by those semantical statements, that to each expression of the types in question (adjectives like "red", numerals like "five", etc.) there is a particular real entity to which the expression stands in the relation of designation. This belief is rejected as incompatible with the basic principles of empiricism or of scientific thinking. Derogatory labels like "Platonic realism", "hypostatization", or " 'Fido'-Fido principle" are attached to it. The latter is the name given by Gilbert Ryle [Meaning] to the criticized belief, which, in his view, arises by a naïve inference of analogy: just as there is an entity well known to me, viz. my dog Fido, which is designated by the name "Fido", thus there must be for every meaningful expression a particular entity to which it stands in the relation of designation or naming, i.e., the relation exemplified by "Fido"-Fido. The belief criticized is thus a case of hypostatization, i.e., of treating as names expressions which

[7] See [I]; *Meaning and Necessity* (Chicago, 1947). The distinction I have drawn in the latter book between the method of the name-relation and the method of intension and extension is not essential for our present discussion. The term "designation" is used in the present article in a neutral way; it may be understood as referring to the name-relation or to the intension-relation or to the extension-relation or to any similar relations used in other semantical methods.

are not names. While "Fido" is a name, expressions like "red", "five", etc., are said not to be names, not to designate anything.

Our previous discussion concerning the acceptance of frameworks enables us now to clarify the situation with respect to abstract entities as designata. Let us take as an example the statement:

(*a*) " 'Five' designates a number".

The formulation of this statement presupposes that our language L contains the forms of expressions which we have called the framework of numbers, in particular, numerical variables and the general term "number". If L contains these forms, the following is an analytic statement in L:

(*b*) "Five is a number".

Further, to make the statement (*a*) possible, L must contain an expression like "designates" or "is a name of" for the semantical relation of designation. If suitable rules for this term are laid down, the following is likewise analytic:

(*c*) " 'Five' designates five".

(Generally speaking, any expression of the form " '. . .' designates . . ." is an analytic statement provided the term ". . ." is a constant in an accepted framework. If the latter condition is not fulfilled, the expression is not a statement.) Since (*a*) follows from (*c*) and (*b*), (*a*) is likewise analytic.

Thus it is clear that *if* someone accepts the framework of numbers, then he must acknowledge (*c*) and (*b*) and hence (*a*) as true statements. Generally speaking, if someone accepts a framework for a certain kind of entities, then he is bound to admit the entities as possible designata. Thus the question of the admissibility of entities of a certain type or of abstract entities in general as designata is reduced to the question of the acceptability of the linguistic framework for those entities. Both the nominalistic critics, who refuse the status of designators or names to expressions like "red", "five", etc., because they deny the existence of abstract entities, and the skeptics, who express doubts concerning the existence and demand evidence for it, treat the question of existence as a theoretical question. They do, of course, not mean the internal question; the affirmative answer to *this* question is analytic and trivial and too obvious for doubt or denial, as we have seen. Their doubts refer rather to the system of entities itself; hence they mean the external question. They believe that only after making sure that there really is a system of entities of the kind in question are we justified in accepting the framework by incorporating the linguistic forms into our language. However, we have seen that the external question is not a theoretical question but rather the practical question whether or not to accept those linguistic forms. This acceptance is not in need of a theoretical justification (except with respect to expediency and fruitfulness), because it does not imply a belief or assertion. Ryle says that the "Fido"-Fido principle is "a grotesque theory". Grotesque or not, Ryle is wrong in calling it a theory. It is rather the practical decision to accept certain frameworks. Maybe Ryle is historically right with respect to those whom he mentions as previous representatives of the principle, viz. John Stuart Mill, Frege, and Russell. If these philosophers regarded the acceptance of a system of entities as a theory, an assertion, they were victims of the same old, metaphysical confusion. But it is certainly wrong to regard *my* semantical method as involving a belief in the reality of abstract entities, since I reject a thesis of this kind as a metaphysical pseudo-statement.

The critics of the use of abstract entities in semantics overlook the fundamental difference between the acceptance of a system of entities and an internal assertion, e.g., an assertion that there are elephants or electrons or prime numbers greater than a million. Whoever makes an internal assertion is certainly obliged to justify it by providing evidence, empirical evidence in the case of electrons, logical proof in the case of the prime numbers. The demand for a theoretical justification, correct in the case of internal assertions, is sometimes wrongly applied to the acceptance of a system of entities. Thus, for example, Ernest Nagel in [Review C.] asks for "evidence relevant for affirming with warrant that there are such entities as infinitesimals or propositions". He characterizes the evidence required in these cases—in distinction to the empirical evidence in the case of electrons—as "in the broad sense logical and dialectical". Beyond this no hint is given as to what might be regarded as relevant evidence. Some nominalists regard the acceptance of abstract entities as a kind of superstition or myth, populating the world with fictitious or at least dubious entities, analogous to the belief in centaurs or demons. This shows again the confusion mentioned, because a superstition or myth is a false (or dubious) internal statement.

Let us take as example the natural numbers as cardinal numbers, i.e., in contexts like "Here are three books". The linguistic forms of the framework of numbers, including variables and the general term "number", are generally used in our common language of communication; and it is easy to formulate explicit rules for their use. Thus the logical characteristics of this framework are sufficiently clear (while many internal questions, i.e., arithmetical questions, are, of course, still open). In spite of this, the controversy concerning the external question of the ontological reality of the system of numbers continues. Suppose that one philosopher says: "I believe that there are numbers as real entities. This gives me the right to use the linguistic forms of the numerical framework and to make semantical statements about numbers as designata of numerals". His nominalistic opponent replies: "You are wrong; there are no numbers. The numerals may still be used as meaningful expressions. But they are not names, there are no entities designated by them. Therefore the word "number" and numerical variables must not be used (unless a way were found to introduce them as merely abbreviating devices, a way of translating them into the nominalistic thing language)." I cannot think of any possible evidence that would be regarded as relevant by both philosophers, and therefore, if actually found, would decide the controversy or at least make one of the opposite theses more probable than the other. (To construe the numbers as classes or properties of the second level, according to the Frege-Russell method, does, of course, not solve the controversy, because the first philosopher would affirm and the second deny the existence of the system of classes or properties of the second level.) Therefore I feel compelled to regard the external question as a pseudo-question, until both parties to the controversy offer a common interpretation of the question as a cognitive question; this would involve an indication of possible evidence regarded as relevant by both sides.

There is a particular kind of misinterpretation of the acceptance of abstract entities in various fields of science and in semantics, that needs to be cleared up. Certain early British empiricists (e.g., Berkeley and Hume) denied the existence of abstract entities on the ground that immediate experience presents us only with particulars, not with universals, e.g., with this red patch, but not with Redness or Color-in-General; with this scalene triangle, but not with Scalene Triangularity or Triangularity-in-General. Only entities belonging to a type of which examples were to be found within immediate experience could be accepted as ultimate constituents of reality. Thus, according to this way of thinking, the existence of abstract entities could be asserted only if one could show either that some abstract entities fall within the given, or that abstract entities can be defined in terms of the types of entity which are given. Since these empiricists found no abstract entities within the realm of sense-data, they either denied their existence, or else made a futile attempt to define universals in terms of particulars. Some contemporary philosophers, especially English philosophers following Bertrand Russell, think in basically similar terms. They emphasize a distinction between the data (that which is immediately given in consciousness, e.g., sense-data, immediately past experiences, etc.) and the constructs based on the data. Existence or reality is ascribed only to the data; the constructs are not real entities; the corresponding linguistic expressions are merely ways of speech not actually designating anything (reminiscent of the nominalists' *flatus vocis*). We shal lnot criticize here this general conception. (As far as it is a principle of accepting certain entities and not accepting others, leaving aside any ontological, phenomenalistic and nominalistic pseudo-statements, there cannot be any theoretical objection to it.) But if this conception leads to the view that other philosophers or scientists who accept abstract entities thereby assert or imply their occurrence as immediate data, then such a view must be rejected as a misinterpretation. References to space-time points, the electromagnetic field, or electrons in physics, to real or complex numbers and their functions in mathematics, to the excitatory potential or unconscious complexes in psychology, to an inflationary trend in economics, and the like, do not imply the assertion that entities of these kinds occur as immediate data. And the same holds for references to abstract entities as designata in semantics. Some of the criticisms by English philosophers against such references give the impression that, probably due to the misinterpretation just indicated, they accuse the semanticist not so much of bad metaphysics (as some nominalists would do) but of bad psychology. The fact that they regard a semantical method involving abstract entities not merely as doubtful and perhaps wrong, but as manifestly absurd, preposterous and grotesque, and that they show a deep horror and indignation against this method, is perhaps to be explained by a misinterpretation of the kind described. In fact, of course, the semanticist does not in the least assert or imply that the abstract entities to which he refers can be experienced as immediately given either by sensation or by a kind of rational intuition. An assertion of this kind would indeed be very dubious psychology. The psychological question as to which kinds of entities do and which do not occur as immediate data is entirely irrelevant for semantics, just as it is for physics, mathematics, economics, etc., with respect to the examples mentioned above.[8]

5. Conclusion

For those who want to develop or use semantical methods, the decisive question is not the alleged ontological question of the existence of abstract entities but rather the question whether the use of abstract linguistic

[8] Wilfrid Sellars ("Acquaintance and Description Again", in *Journal of Philos.*, 46 (1949), 496–504; see pp. 502 f.) analyzes clearly the roots of the mistake "of taking the designation relation of semantic theory to be a reconstruction of *being present to an experience*".

forms or, in technical terms, the use of variables beyond those for things (or phenomenal data), is expedient and fruitful for the purposes for which semantical analyses are made, viz. the analysis, interpretation, clarification, or construction of languages of communication, especially languages of science. This question is here neither decided nor even discussed. It is not a question simply of yes or no, but a matter of degree. Among those philosophers who have carried out semantical analyses and thought about suitable tools for this work, beginning with Plato and Aristotle and, in a more technical way on the basis of modern logic, with C. S. Peirce and Frege, a great majority accepted abstract entities. This does, of course, not prove the case. After all, semantics in the technical sense is still in the initial phases of its development, and we must be prepared for possible fundamental changes in methods. Let us therefore admit that the nominalistic critics may possibly be right. But if so, they will have to offer better arguments than they have so far. Appeal to ontological insight will not carry much weight. The critics will have to show that it is possible to construct a semantical method which avoids all references to abstract entities and achieves by simpler means essentially the same results as the other methods.

The acceptance or rejection of abstract linguistic forms, just as the acceptance or rejection of any other linguistic forms in any branch of science, will finally be decided by their efficiency as instruments, the ratio of the results achieved to the amount and complexity of the efforts required. To decree dogmatic prohibitions of certain linguistic forms instead of testing them by their success or failure in practical use, is worse than futile; it is positively harmful because it may obstruct scientific progress. The history of science shows examples of such prohibitions based on prejudices deriving from religious, mythological, metaphysical, or other irrational sources, which slowed up the developments for shorter or longer periods of time. Let us learn from the lessons of history. Let us grant to those who work in any special field of investigation the freedom to use any form of expression which seems useful to them; the work in the field will sooner or later lead to the elimination of those forms which have no useful function. *Let us be cautious in making assertions and critical in examining them, but tolerant in permitting linguistic forms.*

B. MEANING POSTULATES

1. The Problem of Truth Based upon Meaning

Philosophers have often distinguished two kinds of truth: the truth of some statements is logical, necessary, based upon meaning, while that of other statements is empirical, contingent, dependent upon the facts of the world. The following two statements belong to the first kind:

(1) 'Fido is black or Fido is not black'

(2) 'If Jack is a bachelor, then he is not married'

In either case it is sufficient to understand the statement in order to establish its truth; knowledge of (extra-linguistic) facts is not involved. However, there is a difference. To ascertain the truth of (1), only the meanings of the logical particles ('is', 'or', 'not') are required; the meanings of the descriptive (i.e., nonlogical) words ('Fido', 'black') are irrelevant (except that they must belong to suitable types). For (2), on the other hand, the meanings of some descriptive words are involved, viz., those of 'bachelor' and 'married'.

Quine[1] has recently emphasized the distinction; he uses the term 'analytic' for the wider kind of statement to which both examples belong, and 'logically true' for the narrower kind to which (1) belongs but not (2). I shall likewise use these two terms for the explicanda. But I do not share Quine's skepticism; he is doubtful whether an explication of analyticity, especially one in semantics, is possible, and even whether there is a sufficiently clear explicandum, especially with respect to natural languages.

It is the purpose of this paper to describe a way of explicating the concept of analyticity, i.e., truth based upon meaning, in the framework of a semantical system, by using what we shall call *meaning postulates*. This simple way does not involve any new idea; it is rather suggested by a common-sense reflection. It will be shown in this paper how the definitions of some concepts fundamental for deductive and inductive logic can be reformulated in terms of postulates.[2]

Our explication, as mentioned above, will refer to semantical language

[1] W. V. Quine [Dogmas], especially pp. 23 f.

[2] This paper presupposes the explication of logical truth, which will be indicated in § 2, and that of the distinction between logical and descriptive constants (compare [I], § 13). Our present task is only to solve the additional problem involved in the explication of analyticity.

systems, not to natural languages. It shares this character with most of the explications of philosophically important concepts given in modern logic, e.g., Tarski's explication of truth. It seems to me that the problems of explicating concepts of this kind for natural languages are of an entirely different nature.[3]

2. Meaning Postulates

Our discussion refers to a semantical language system $\mathfrak{L}$ of the following kind. $\mathfrak{L}$ contains the customary connectives, individual variables with quantifiers, and as descriptive signs individual constants ('*a*', '*b*', etc.) and primitive descriptive predicates (among them '*B*', '*M*', '*R*', and '*Bl*', for the properties Bachelor, Married, Raven, and Black, respectively). The following statements in $\mathfrak{L}$ correspond to the two earlier examples:

(3) '$Bl\ a \vee \sim Bl\ a$'

(4) '$B\ b \supset \sim M\ b$'

Suppose that the customary truth-tables for the connectives are laid down for $\mathfrak{L}$ (in the form of rules of truth or satisfaction) but that *no* rules of designation for the descriptive constants are given (hence the meanings of the four predicates mentioned above are *not* incorporated into the system). Before we state meaning postulates, let us see what can be done without them, on the basis of semantical rules of the customary kinds. First let us define the L-truth of a sentence $\mathfrak{S}_i$ of $\mathfrak{L}$ as an explicatum for logical truth (in the narrow sense). We may use as definiens any one of the subsequent four formulations (5a) to (5d); they are equivalent to one another (provided they are applicable to $\mathfrak{L}$). Insertions in square brackets refer to example (3).

(5a) The open logical formula corresponding to $\mathfrak{S}_i$ [e.g., '$fx \vee \sim fx$'] is universally valid (i.e., satisfied by all values of the free variables). (Here it is presupposed that $\mathfrak{L}$ contains corresponding variables for all descriptive constants.)

(5b) The universal logical statement corresponding to $\mathfrak{S}_i$ [e.g., '$(f)(x)(fx \vee \sim fx)$'] is true. (Here it is presupposed that $\mathfrak{L}$ has variables with quantifiers corresponding to all descriptive constants.)

[3] The great difficulties and complications of any attempt to explicate logical concepts for natural languages have been clearly explained by Benson Mates in [Analytic] and by Richard Martin in [Analytic]. Both articles offer strong arguments against the view held by Quine [Dogmas] and Morton G. White [Analytic] that there is no clear distinction between analytic and synthetic.

(5c) $\mathfrak{S}_i$ is satisfied by all values of the descriptive constants occurring. [The ranges of values for '*Bl*' and '*a*' here are the same as those for '*f*' and '*x*', respectively, in (5a).]

(5d) $\mathfrak{S}_i$ holds in all state-descriptions. (A state-description is a conjunction containing for every atomic statement either it or its negation but not both, and no other statements. Here it is presupposed that $\mathfrak{L}$ contains constants for all values of its variables and, in particular, individual constants for all individuals of the universe of discourse.)

Each of these formulations presupposes, of course, that rules for the system $\mathfrak{L}$ are given which determine the concepts involved, e.g., rules of formation (determining the forms of open formulas and statements, i.e., closed formulas), rules for the range of values of all variables and for (5c) also analogous rules for the range of values for all descriptive constants,[4] and for (5d) rules determining those state-descriptions in which any given statement holds. Form (5d) is quite convenient if $\mathfrak{L}$ has the required form. Form (5c) imposes the least restrictions on $\mathfrak{L}$.

The other concepts can easily be defined on the basis of L-truth. Thus L-falsity, L-implication, and L-equivalence may be defined by the L-truth of $\sim\mathfrak{S}_i$, $\mathfrak{S}_i \supset \mathfrak{S}_j$, and $\mathfrak{S}_i \equiv \mathfrak{S}_j$, respectively.

The definition of L-truth in $\mathfrak{L}$, in any one of the four alternative forms, covers example (3) but obviously not (4). In order to provide for (4), we lay down the following *meaning postulate:*

$$(\mathrm{P}_1)\ `(x)(Bx \supset \sim Mx)'$$

Even now we do not give rules of designation for '*B*' and '*M*'. They are not necessary for the explication of analyticity, but only for that of factual (synthetic) truth. But postulate P_1 states as much about the meanings of '*B*' and '*M*' as is essential for analyticity, viz., the incompatibility of the two properties. If logical relations (e.g., logical implication or incompatibility) hold between the intended meanings of the primitive predicates of a system, then the explication of analyticity requires that postulates for all such relations are laid down. The term 'postulate' seems suitable for this purpose; it has sometimes been used in a similar sense.[5] (This usage is not the same as the more frequent one according to which 'postulate' is synonymous with 'axiom'.)

Suppose that the author of a system wishes the predicates '*B*' and '*M*' to designate the properties Bachelor and Married, respectively. How does

[4] Compare [Syntax], § 34c.

[5] See, for example, J. Cooley, *Primer of Formal Logic* (1942), p. 153.

he know that these properties are incompatible and that therefore he has to lay down postulate P_1? This is not a matter of knowledge but of decision. His knowledge or belief that the English words 'bachelor' and 'married' are always or usually understood in such a way that they are incompatible may influence his decision if he has the intention to reflect in his system some of the meaning relations of English words. In this particular case, the influence would be relatively clear, but in other cases it would be much less so.

Suppose he wishes the predicates '*Bl*' and '*R*' to correspond to the words 'black' and 'raven'. While the meaning of 'black' is fairly clear, that of 'raven' is rather vague in the everyday language. There is no point for him to make an elaborate study, based either on introspection or on statistical investigation of common usage, in order to find out whether 'raven' always or mostly entails 'black'. It is rather his task to make up his mind whether he wishes the predicates '*R*' and '*Bl*' of his system to be used in such a way that the first logically entails the second. If so, he has to add the postulate

$$(P_2)\ `(x)(Rx \supset Bl\,x)'$$

to the system, otherwise not.

Suppose the meaning of '*Bl*', viz., Black, is clear to him. Then the two procedures between which he has to choose may be formulated as follows: (1) he wishes to give to '*R*' a meaning so strong that it cannot possibly be predicated of any non-black thing; (2) he gives to '*R*' a certain (weaker) meaning; although he may believe that all things to which '*R*' applies are black so that he would be greatly surprised if he found one that was not black, the intended meaning of '*R*' does not by itself rule out such an occurrence. Thus we see that it cannot be the task of the logician to prescribe to those who construct systems what postulates they ought to take. They are free to choose their postulates, guided not by their beliefs concerning facts of the world but by their intentions with respect to the meanings, i.e., the ways of use of the descriptive constants.

Suppose that certain meaning postulates have been accepted for the system $\mathfrak{L}$. Let $\mathfrak{P}$ be their conjunction. Then the concept of analyticity, which applies to both examples (3) and (4), can now be explicated. We shall use for the explicatum the term 'L-true with respect to $\mathfrak{P}$' and define it as follows:

(6) A statement $\mathfrak{S}_i$ in $\mathfrak{L}$ is L-true with respect to $\mathfrak{P} =_{\mathrm{Df}} \mathfrak{S}_i$ is L-implied by $\mathfrak{P}$ (in $\mathfrak{L}$).[6]

[6] The term 'L-true with respect to $\mathfrak{P}$' is simply a special case of the relative L-terms which I have used elsewhere; see [Probability], D20-2.

The definiens could, of course, also be formulated as "$\mathfrak{P} \supset \mathfrak{S}_i$ is L-true (in $\mathfrak{L}$)" or "$\mathfrak{S}_i$ holds in all state-descriptions in which $\mathfrak{P}$ holds" (the latter presupposes that L-truth in $\mathfrak{L}$ is defined by (5d)).

The definitions of the other L-concepts with respect to $\mathfrak{P}$ in terms of L-truth with respect to $\mathfrak{P}$ are analogous to the earlier definitions and therefore need not be stated here. The following theorem can be seen to result immediately:

(7) Each of the following conditions (a) to (d) is a sufficient and necessary condition for $\mathfrak{S}_i$ L-implying $\mathfrak{S}_j$ with respect to $\mathfrak{P}$:
 (a) $\mathfrak{P}$ L-implies $\mathfrak{S}_i \supset \mathfrak{S}_j$
 (b) $\mathfrak{P} \supset (\mathfrak{S}_i \supset \mathfrak{S}_j)$ is L-true
 (c) $P \bullet \mathfrak{S}_i \supset \mathfrak{S}_j$ is L-true
 (d) $P \bullet \mathfrak{S}_i$ L-implies $\mathfrak{S}_j$

An alternative way, differing merely in the form of systematization but leading to the same results, would be as follows. Let $\mathfrak{L}$ be the original system without meaning postulates. The system $\mathfrak{L}'$ is constructed out of $\mathfrak{L}$ by adding the meaning postulates $\mathfrak{P}$. Then we define:

(8) $\mathfrak{S}_i$ is L-true in $\mathfrak{L}' =_{\mathrm{Df}} \mathfrak{S}_i$ is L-implied by $\mathfrak{P}$ in $\mathfrak{L}$.

L-truth in $\mathfrak{L}'$ is then the explicatum for analyticity.

If L-truth in $\mathfrak{L}$ is defined by (5d), then the following definitions could take the place of (8):

(9) The state-descriptions in $\mathfrak{L}' =_{\mathrm{Df}}$ those state-descriptions in $\mathfrak{L}$ in which $\mathfrak{P}$ holds.

(10) $\mathfrak{S}_i$ is L-true in $\mathfrak{L}' =_{\mathrm{Df}} \mathfrak{S}_i$ holds in every state-description in $\mathfrak{L}'$.

The other L-concepts in $\mathfrak{L}'$ are then defined in terms of L-truth in $\mathfrak{L}'$ in the same way as before. If, for example, $\mathfrak{P}$ contains the postulates P_1 and P_2 mentioned earlier, then the following results would hold in $\mathfrak{L}'$: '$B\,b \supset \sim M\,b$' and '$R\,a \supset Bl\,a$' are L-true; '$B\,b \bullet M\,b$' and '$R\,a \bullet \sim Bl\,a$' are L-false; '$B\,b$' L-implies '$\sim M\,b$', and '$R\,a$' L-implies '$Bl\,a$'; '$R\,a \bullet Bl\,a$' is L-equivalent to '$R\,a$'.

3. Meaning Postulates for Relations

Suppose that among the primitive predicates there are also some with two or more arguments designating two- or more-place relations, and that one of these predicates possesses some structural properties in virtue of its meaning. For example, let '*W*' be a primitive predicate designating the relation Warmer. Then '*W*' is transitive, irreflexive, and hence asymmetric in virtue of its meaning. Therefore the statements '$Wab \bullet Wbc \bullet \sim Wac$', '$Wab \bullet Wba$', and '$Waa$' are false due to their meanings. The same holds for state-descriptions which contain one of these statements as subconjunctions; hence they do not represent possible cases. This difficulty was discovered by John G. Kemeny[7] and Yehoshua Bar-Hillel[8] independently. It is more serious than that due to logical dependencies between two or more one-place predicates, as it cannot be avoided by simply replacing dependent by independent predicates with the same expressive power.

There are two ways of overcoming the difficulty. The first, which maintains the requirement of the logical independence of all atomic statements, consists in avoiding primitive relations entirely or at least those of the customary kinds.[9]

The second way abandons the requirement of independence. It admits dependent primitives including relational ones, but restricts state-descriptions to those which represent possible cases, by stating meaning postulates or other equivalent rules. This way was first proposed by Kemeny.[10] In comparison with the first way, the second has the disadvantage of needing a new semantical concept (either 'directly L-true', i.e., 'meaning postulate', or 'directly L-false' in an alternative procedure), defined by enumeration in each semantical system or taken as primitive in general semantics. Another disadvantage is the more complicated form of theorems and computations of values of the degree of confirmation in inductive logic. For these reasons, Bar-Hillel and I previously did not pursue the second way any further.[11] On the other hand, it has the advantage of giving more freedom in the choice of primitives.

In the previous example of the predicate 'W', we could lay down the following postulates (a) for transitivity and (b) for irreflexivity; then the statement (c) of asymmetry is L-true with respect to these two postulates:

(11) (a) '$(x)(y)(z)\ (Wxy \bullet Wyz \supset Wxz)$'

(b) '$(x) \sim Wxx$'

(c) '$(x)(y)\ (Wxy \supset \sim Wyx)$'

[7] J. G. Kemeny, review of [Probability] in *Journal of Symbolic Logic*, 16 (1951), 205–7.

[8] Y. Bar-Hillel, "A Note on State-Descriptions", *Philosophical Studies*, 2 (1951), 72–75. Compare my reply, "The Problem of Relations in Inductive Logic", *ibid.*, 75–80.

[9] Some possibilities of this are outlined in my paper mentioned in the preceding footnote.

[10] See footnote 7. The procedure was carried out by Kemeny in "Extension of the Methods of Inductive Logic", *Philosophical Studies*, 3 (1952), 38–42, and in "A Logical Measure Function", *Journal of Symbolic Logic*, 18 (1953), 289–308. (These two articles were not known to me when I wrote the present paper.)

[11] See Bar-Hillel, *op. cit.*, p. 74, "the third possibility".

If we admit the form of semantical rules which we have called meaning postulates, we find that other customary kinds of rules may be construed as special kinds of meaning postulates. This holds, for example, for explicit definitions (if written as statements in the object language with '$\equiv$' or '$=$') and for contextual definitions. Likewise, the two or more formulas of a so-called recursive definition of an arithmetical functor may be regarded as meaning postulates. In this case, the label 'postulate' is perhaps even more appropriate than the customary one of 'definition'. The formulas serve not merely for an introduction of an abbreviating notation, since the new functor is not eliminable in all contexts. Further, the reduction-sentences which I proposed earlier for the introduction of disposition predicates[12] may be construed as meaning postulates.

[A bilateral reduction-sentence '$(x)\ [Q_1x \supset (Q_3x \equiv Q_2x)]$' for '$Q_3$' may simply be taken as a postulate, since it has no synthetic consequences in terms of the original predicates 'Q_1' and 'Q_2'. This is, however, in general not possible for the formulas of a reduction pair, e.g., '$(x)\ [Q_1x \supset (Q_2x \supset Q_3x)]$' ($\mathfrak{S}_1$) and '$(x)\ [Q_4x \supset (Q_5x \supset \sim Q_3x)]$' ($\mathfrak{S}_2$), since they together imply the synthetic statement '$(x) \sim (Q_1x \bullet Q_2x \bullet Q_4x \bullet Q_5x)$' ($\mathfrak{S}_3$). Here, we must take as postulate the weaker statement $\mathfrak{S}_3 \supset \mathfrak{S}_1 \bullet \mathfrak{S}_2$, which has no synthetic consequences.]

4. Meaning Postulates in Inductive Logic

A few brief remarks may be made here concerning the consequences of the use of meaning postulates for inductive logic. Let $\mathfrak{m}$ be any regular measure-function for the system $\mathfrak{L}$, and $\mathfrak{c}$ be the confirmation-function based upon $\mathfrak{m}$ (i.e., $\mathfrak{c}(h,e) = \mathfrak{m}(e \bullet h)/\mathfrak{m}(e)$). Let $\mathfrak{m}'$ be a function for the state-descriptions in $\mathfrak{L}$ fulfilling the following three conditions:

(12) (a) For any state-description k in $\mathfrak{L}$ in which $\mathfrak{P}$ does not hold, $\mathfrak{m}'(k) = 0$.

(b) For any state-description k in $\mathfrak{L}$ in which $\mathfrak{P}$ holds, $\mathfrak{m}'(k)$ is proportional to $\mathfrak{m}(k)$; say, $\mathfrak{m}'(k) = K\mathfrak{m}(k)$.

(c) The sum of the $\mathfrak{m}'$-values for all state-descriptions in $\mathfrak{L}$ is 1.

It is easily seen that, for any regular function $\mathfrak{m}$, there is one and only one function $\mathfrak{m}'$ of this kind. We find from (b) and (c) that K must be $1/\mathfrak{m}(\mathfrak{P})$. Since for the state-descriptions in $\mathfrak{L}'$, $\mathfrak{m}'$ has positive values (according to (9) and (12)(b)) whose sum is 1, $\mathfrak{m}'$ may be regarded as the regular function for $\mathfrak{L}'$ corresponding to $\mathfrak{m}$ for $\mathfrak{L}$.

[12] "Testability and Meaning", *Philosophy of Science*, Vols. 3 and 4 (1936 and 1937); reprinted by Whitlock's Inc., New Haven, Connecticut, 1950; see §§ 8–10.

Let $\mathfrak{m}'$ be applied to other statements in the customary way, and let the function $\mathfrak{c}'$ for $\mathfrak{L}'$ be based upon $\mathfrak{m}'$ (i.e., $\mathfrak{c}'(h,e) = \mathfrak{m}'(e \bullet h)/\mathfrak{m}'(e)$). Then $\mathfrak{c}'$ may be regarded as the regular confirmation-function for $\mathfrak{L}'$ corresponding to $\mathfrak{c}$ for $\mathfrak{L}$. The following results are easily obtained:

(13) For any state-description k in $\mathfrak{L}'$ (which is a state-description in $\mathfrak{L}$ in which $\mathfrak{P}$ holds), $\mathfrak{m}'(k) = \mathfrak{m}(k)/\mathfrak{m}(\mathfrak{P})$.

(14) For any statement j, $\mathfrak{m}'(j) = \mathfrak{m}(\mathfrak{P} \bullet j)/\mathfrak{m}(\mathfrak{P}) = \mathfrak{c}(j,\mathfrak{P})$.

(15) For any statements h and e, where e is not L-false in $\mathfrak{L}'$ (and hence $\mathfrak{P} \bullet e$ is not L-false in $\mathfrak{L}$), $\mathfrak{c}'(h,e) = \mathfrak{m}'(e \bullet h)/\mathfrak{m}'(e)$ $= \mathfrak{m}(\mathfrak{P} \bullet e \bullet h)/\mathfrak{m}(\mathfrak{P} \bullet e) = \mathfrak{c}(h,\mathfrak{P} \bullet e)$.

We see that the degree of confirmation in a system with postulates $\mathfrak{P}$ has in each case the same value as that obtained in the original system by adding $\mathfrak{P}$ to the evidence. This is analogous to the earlier result, according to which $\mathfrak{S}_i$ L-implies $\mathfrak{S}_j$ in $\mathfrak{L}'$ if and only if $\mathfrak{S}_i \bullet \mathfrak{P}$ L-implies $\mathfrak{S}_j$ in $\mathfrak{L}$ (compare (7)(d)). With the help of (15), general theorems concerning regular confirmation functions for systems with meaning postulates can easily be obtained from the known theorems for systems without postulates. However, if primitive relations occur and postulates are laid down for structural properties of these relations, then the computation of values of a particular function, e.g., $\mathfrak{c}^*$, will in many cases become even more complicated than in a system with the same primitives but without postulates.

C. ON BELIEF-SENTENCES

REPLY TO ALONZO CHURCH

Church's paper [Belief] raises objections against the explication of belief-sentences which I had proposed in my book *Meaning and Necessity*. The first part of Church's paper does not apply to my analysis because the latter does not refer to historically given languages, but rather to semantical systems, which are defined by their rules. Thus only the objection stated in Church's last paragraph applies. This objection is correct, but it can be met by a modification in my explication of belief-sentences, suggested by Putnam [Synonymity]. I shall not discuss this point here, because at present I am inclined, for general reasons, to make a more radical change in that explication.

It seems best to reconstruct the language of science in such a way that terms like 'temperature' in physics or 'anger' or 'belief' in psychology are introduced as theoretical constructs rather than as intervening variables of the observation language. This means that a sentence containing a term of this kind can neither be translated into a sentence of the language of observables nor deduced from such sentences, but at best inferred with high probability. I think this view is at present shared by most logical empiricists; it has been expounded with great clarity and convincing arguments by Feigl[1] and Hempel.[2]

In application to belief-sentences, this means that a sentence like

(i) John believes that the earth is round,

is to be interpreted in such a way that it can be inferred from a suitable sentence describing John's behavior at best with probability, but not with certainty, e.g., from

(ii) John makes an affirmative response to "the earth is round" as an English sentence.

When I wrote my book, I had already developed the general view mentioned above, concerning the nature of sentences in physics and psychol-

[1] H. Feigl, "Existential Hypotheses", *Phil. of Science*, 17 (1950), 35–62; "Principles and Problems of Theory Construction in Psychology", in *Current Trends in Psychological Theory* (University of Pittsburgh Press, 1951), pp. 179–213.

[2] C. G. Hempel, *Fundamentals of Concept Formation in Empirical Science*, Encycl. Unified Science, Vol. II, No. 7 (1952).

ogy. However, I believed then erroneously that for the intended semantical analysis the simplification involved in taking a response as a conclusive evidence for a belief, would not essentially change the problem. It seems that Benson Mates was the first to see the difficulty involved, although not its solution. He pointed out ([Synonymity], p. 215) that any two different sentences, no matter how similar, could evoke different psychological responses. He argued that therefore my explication of synonymity, and likewise any other one, would lead to difficulties, e.g., in the case of the following two sentences:

(iii) Whoever believes that *D*, believes that *D*,

(iv) Whoever believes that *D*, believes that *D′*,

where '*D*' and '*D′*' are abbreviations for two different but synonymous sentences. Then (iii) and (iv) would themselves be synonymous. However, while (iii) is certainly true and beyond doubt, (iv) may be false or, at least, it is conceivable that somebody may doubt it. This is indeed a serious difficulty, but only as long as we regard an affirmative response to '*D*' as a conclusive indication of belief in *D*.

Church pointed out to me that Mates' paradoxical result concerning (iii) and (iv) disappears if we give up that view. We may then take (iv) as logically true, just like (iii). If somebody responds affirmatively to '*D*', but negatively to '*D′*', we shall merely conclude that one of his responses is non-indicative, perhaps due to his momentary confusion.

While I agree with Church in this point, there remains a divergence of view with respect to the question of the best form for belief-sentences in a formalized language of science. One form uses indirect discourse in analogy to the form (i) of ordinary language. The other form avoids indirect discourse; here a belief-sentence does not like (i) contain a partial sentence expressing the content of the belief, but instead the name of such a sentence, for example:

(v) John has the relation *B* to "the earth is round" as a sentence in English.

It is to be noted that, according to the new interpretation explained above, (v) is not deducible from (ii) but is merely confirmed by it to some degree. '*B*' is a theoretical construct, not definable in terms of overt behavior, be it linguistic or non-linguistic. The rules for '*B*' would be such that (v) does not imply that John knows English or any language whatsoever. On the other hand, the reference to an English sentence in (v) may be replaced by a reference to any other synonymous sentence in any language; e.g., (v) is taken to be L-equivalent with:

(vi) John has the relation *B* to "die Erde ist rund" as a sentence in German.

As an explication of synonymity we may use here the relation of intensional isomorphism as proposed in my book; it holds, if the two expressions are constructed in the same way out of signs with the same intensions; as an alternative, a slightly stronger relation, suggested by Putnam, may be used, which requires that the two expressions have, in addition, the same syntactical structure.

Church entertains the view that a belief must be construed as a relation between a person and a proposition, not a sentence, and that therefore only the first form, like (i), is adequate, not the second, like (v). I do not reject the first form, but regard both forms as possible. I do not think that the arguments offered by Church so far show the impossibility of the second form. Both forms must be further investigated before we can decide which one is preferable. It must be admitted that the second form has certain disadvantages; it abolishes the customary and convenient device of indirect discourse, it uses the metalanguage, and it becomes cumbersome in cases of iteration (e.g., "James asserts that John believes that . . ." would be replaced by a sentence about a sentence about a sentence). The main disadvantage of the first form is the complexity of the logical structure of the language, whereas the language for the second form may be extensional and therefore very simple. The introduction of logical modalities produces already considerable complications, but the use of indirect discourse increases them still more. The greatest complexity would result from the use of the Frege-Church method, according to which an expression has infinitely many senses depending upon the text (see my book, pp. 129 ff.). Church believes that these complications are inevitable, but I am not convinced of it. I regard it as possible to construct a language of the first form in such a way that every expression has always the same sense and that therefore two expressions which fulfil a certain criterion of synonymity are synonymous in any context, including contexts of simple or iterated indirect discourse. But many more investigations and tentative constructions of languages will have to be made before we can see the whole situation clearly and make a well-founded decision as to the choice of the language form.

D. MEANING AND SYNONYMY IN NATURAL LANGUAGES

1. Meaning Analysis in Pragmatics and Semantics

The analysis of meanings of expressions occurs in two fundamentally different forms. The first belongs to *pragmatics*, that is, the empirical investigation of historically given *natural languages*. This kind of analysis has long been carried out by linguists and philosophers, especially analytic philosophers. The second form was developed only recently in the field of symbolic logic; this form belongs to *semantics* (here understood in the sense of pure semantics, while descriptive semantics may be regarded as part of pragmatics), that is, the study of constructed *language systems* given by their rules.

The theory of the relations between a language—either a natural language or a language system—and what language is about may be divided into two parts which I call the theory of extension and the theory of intension, respectively.[1] The first deals with concepts like denoting, naming, extension, truth, and related ones. (For example, the word 'blau' in German, and likewise the predicate '*B*' in a symbolic language system if a rule assigns to it the same meaning, denote any object that is blue; its extension is the class of all blue objects; 'der Mond' is a name of the moon; the sentence 'der Mond ist blau' is true if and only if the moon is blue.) The theory of intension deals with concepts like intension, synonymy, analyticity, and related ones; for our present discussion let us call them "*intension concepts*". (I use 'intension' as a technical term for the meaning of an expression or, more specifically, for its designative meaning component; see below. For example, the intension of 'blau' in German is the property of being blue; two predicates are synonymous if and only if they have the same intension; a sentence is analytic if it is true by virtue of the intensions of the expressions occurring in it.)

From a systematic point of view, the description of a language may well begin with the theory of intension and then build the theory of extension on its basis. By learning the theory of intension of a language, say German, we learn the intensions of the words and phrases and finally of the sen-

[1] This distinction is closely related to that between radical concepts and L-concepts which I made in [I]. The contrast between extension and intension is the basis of the semantical method which I developed in *Meaning and Necessity*. Quine calls the two theories "theory of reference" and "theory of meaning", respectively.

tences. Thus the theory of intension of a given language L enables us to *understand* the sentences of L. On the other hand, we can apply the concepts of the theory of extension of L only if we have, in addition to the knowledge of the theory of intension of L, also sufficient empirical knowledge of the relevant facts. For example, in order to ascertain whether a German word denotes a given object, one must first understand the word, that is, know what is its intension, in other words, know the general condition which an object must fulfil in order to be denoted by this word; and secondly he must investigate the object in question in order to see whether it fulfils the condition or not. On the other hand, if a linguist makes an empirical investigation of a language not previously described, he finds out first that certain objects are denoted by a given word, and later he determines the intension of the word.

Nobody doubts that the pragmatical investigation of natural languages is of greatest importance for an understanding both of the behavior of individuals and of the character and development of whole cultures. On the other hand, I believe with the majority of logicians today that for the special purpose of the development of logic the construction and semantical investigation of language systems is more important. But also for the logician a study of pragmatics may be useful. If he wishes to find out an efficient form for a language system to be used, say, in a branch of empirical science, he might find fruitful suggestions by a study of the natural development of the language of scientists and even of the everyday language. Many of the concepts used today in pure semantics were indeed suggested by corresponding pragmatical concepts which had been used for natural languages by philosophers or linguists, though usually without exact definitions. Those semantical concepts were, in a sense, intended as explicata for the corresponding pragmatical concepts.

In the case of the semantical intension concepts there is an additional motivation for studying the corresponding pragmatical concepts. The reason is that some of the objections raised against these semantical concepts concern, not so much any particular proposed explication, but the question of the very existence of the alleged explicanda. Especially *Quine's* criticism does not concern the formal correctness of the definitions in pure semantics; rather, he doubts whether there are any clear and fruitful corresponding pragmatical concepts which could serve as explicanda. That is the reason why he demands that these pragmatical concepts be shown to be scientifically legitimate by stating empirical, behavioristic criteria for them. If I understand him correctly, he believes that, without this pragmatical substructure, the semantical intension concepts, even if formally correct, are arbitrary and without purpose. I do not think that a semantical concept, in order to be fruitful, must necessarily possess a prior pragmatical counterpart. It is theoretically possible to demonstrate its fruitfulness through its application in the further development of language systems. But this is a slow process. If for a given semantical concept there is already a familiar, though somewhat vague, corresponding pragmatical concept and if we are able to clarify the latter by describing an operational procedure for its application, then this may indeed be a simpler way for refuting the objections and furnish a practical justification at once for both concepts.

The purpose of this paper is to clarify the nature of the pragmatical concept of intension in natural languages and to outline a behavioristic, operational procedure for it. This will give a practical vindication for the semantical intension concepts; ways for defining them, especially analyticity, I have shown in a previous paper [Postulates]. By way of introduction I shall first (in § 2) discuss briefly the pragmatical concepts of denotation and extension; it seems to be generally agreed that they are scientifically legitimate.

2. The Determination of Extensions

We take as example the German language. We imagine that a linguist who does not know anything about this language sets out to study it by observing the linguistic behavior of German-speaking people. More specifically, he studies the German language as used by a given person Karl at a given time. For simplicity, we restrict the discussion in this paper mainly to predicates applicable to observable things, like 'blau' and 'Hund'. It is generally agreed that, on the basis of spontaneous or elicited utterances of a person, the linguist can ascertain whether or not the person is willing to apply a given predicate to a given thing, in other words, whether the predicate denotes the given thing for the person. By collecting results of this kind, the linguist can determine first, the extension of the predicate 'Hund' within a given region for Karl, that is, the class of the things to which Karl is willing to apply the predicate, second, the extension of the contradictory, that is, the class of those things for which Karl denies the application of 'Hund', and, third, the intermediate class of those things for which Karl is not willing either to affirm or to deny the predicate. The size of the third class indicates the degree of vagueness of the predicate 'Hund', if we disregard for simplicity the effect of Karl's ignorance about relevant facts. For certain predicates, e.g., 'Mensch', this third class is relatively very small; the degree of their extensional vagueness is low. On the basis of the determination of the three classes for the predicate 'Hund' within the investigated region, the linguist may make a hypothesis concerning the responses of Karl to things outside of that region, and maybe even a hypothesis concerning the total extension in the universe. The latter hypothesis cannot, of course, be completely verified, but every single instance of it can in principle be tested. On the other hand, it is also generally agreed that this determination of extension involves uncertainty and possible error. But since this holds for all concepts of empirical science, nobody regards this fact as a sufficient reason for rejecting the concepts of the theory of extension. The sources of uncertainty are chiefly the following: first, the linguist's acceptance of the result that a given thing is denoted by 'Hund' for Karl may be erroneous, e.g., due to a misunderstanding or a factual error of Karl's; and, second, the generalization to things which he has not tested suffers, of course, from the uncertainty of all inductive inference.

3. The Determination of Intensions

The purpose of this paper is to defend the thesis that the analysis of intension for a natural language is a scientific procedure, methodologically just as sound as the analysis of extension. To many linguists and philosophers this thesis will appear as a truism. However, some contemporary philosophers, especially Quine[2] and White [Analytic] believe that the pragmatical intension concepts are foggy, mysterious, and not really understandable, and that so far no explications for them have been given. They believe further that, if an explication for one of these concepts is found, it will at best be in the form of a concept of degree. They acknowledge the good scientific status of the pragmatical concepts of the theory of extension. They emphasize that their objection against the intension concepts is based on a point of principle and not on the generally recognized facts of the technical difficulty of linguistic investigations, the inductive uncertainty, and the vagueness of the words of ordinary language. I shall therefore leave aside in my discussion these difficulties, especially the two mentioned at the end of the last section. Thus the question is this: *granted that the linguist can determine the extension of a given predicate, how can he go beyond this and determine also its intension?*

The technical term 'intension', which I use here instead of the ambiguous word 'meaning', is meant to apply only to the cognitive or designative meaning component. I shall not try to define this component. It was men-

[2] W. V. Quine [Logical]; for his criticism of intension concepts see especially Essays II [Dogmas], III, and VII.

tioned earlier that determination of truth presupposes knowledge of meaning (in addition to knowledge of facts); now, cognitive meaning may be roughly characterized as that meaning component which is relevant for the determination of truth. The non-cognitive meaning components, although irrelevant for questions of truth and logic, may still be very important for the psychological effect of a sentence on a listener, e.g., by emphasis, emotional associations, motivational effects.

It must certainly be admitted that the pragmatical determination of intensions involves a new step and therefore a new methodological problem. Let us assume that two linguists, investigating the language of Karl, have reached complete agreement in the determination of the extension of a given predicate in a given region. This means that they agree for every thing in this region, whether or not the predicate in question denotes it for Karl. As long as only these results are given, no matter how large the region is—you may take it, fictitiously, as the whole world, if you like—it is still possible for the linguists to ascribe to the predicate different intensions. For there are more than one and possibly infinitely many properties whose extension within the given region is just the extension determined for the predicate.

Here we come to the core of the controversy. It concerns the nature of a linguist's assignment of one of these properties to the predicate as its intension. This assignment may be made explicit by an entry in the German-English dictionary, conjoining the German predicate with an English phrase. The linguist declares hereby the German predicate to be synonymous with the English phrase. *The intensionalist thesis* in pragmatics, which I am defending, says that the assignment of an intension is an empirical hypothesis which, like any other hypothesis in linguistics, can be tested by observations of language behavior. On the other hand, *the extensionalist thesis* asserts that the assignment of an intension, on the basis of the previously determined extension, is not a question of fact but merely a matter of choice. The thesis holds that the linguist is free to choose any of those properties which fit to the given extension; he may be guided in his choice by a consideration of simplicity, but there is no question of right or wrong. Quine seems to maintain this thesis; he says: "The finished lexicon is a case evidently of *ex pede Herculem*. But there is a difference. In projecting Hercules from the foot we risk error but we may derive comfort from the fact that there is something to be wrong about. In the case of the lexicon, pending some definition of synonymy, we have no stating of the problem; we have nothing for the lexicographer to be right or wrong about." ([Logical], p. 63.)

I shall now plead for the intensionalist thesis. Suppose, for example, that one linguist, after an investigation of Karl's speaking behavior, writes into his dictionary the following:

(1) *Pferd*, horse,

while another linguist writes:

(2) *Pferd*, horse or unicorn.

Since there are no unicorns, the two intensions ascribed to the word 'Pferd' by the two linguists, although different, have the same extension. If the extensionalist thesis were right, there would be no way for empirically deciding between (1) and (2). Since the extension is the same, no response by Karl, affirmative or negative, with respect to any actual thing can make a difference between (1) and (2). But what else is there to investigate for the linguist beyond Karl's responses concerning the application of the predicate to all the cases that can be found? The answer is, he must take into account not only the actual cases, but also possible cases.[3] The most direct way of doing this would be for the linguist to use, in the German questions directed to Karl, modal expressions corresponding to "possible case" or the like. To be sure, these expressions are usually rather ambiguous; but this difficulty can be overcome by giving suitable explanations and examples. I do not think that there is any objection of principle against the use of modal terms. On the other hand, I think that their use is not necessary. The linguist could simply describe for Karl cases, which he knows to be possible, and leave it open whether there is anything satisfying those descriptions or not. He may, for example, describe a unicorn (in German) by something corresponding to the English formulation: "a thing similar to a horse, but having only one horn in the middle of the forehead". Or he may point toward a thing and then describe the intended modification in words, e.g.: "a thing like this one but having one horn in the middle of the forehead". Or, finally, he might just point to a picture representing a unicorn. Then he asks Karl whether he is willing to apply the word 'Pferd' to a thing of this kind. An affirmative or a negative answer will constitute a confirming instance for (2) or (1) respectively. This shows that (1) and (2) are different empirical hypotheses.

[3] Some philosophers have indeed defined the intension of a predicate (or a concept closely related to it) as the class of the possible objects falling under it. For example, C. I. Lewis defines: "The comprehension of a term is the classification of all consistently thinkable things to which the term would correctly apply." I prefer to apply modalities like possibility not to objects but only to intensions, especially to propositions or to properties (kinds). (Compare *Meaning and Necessity*, pp. 66 f.) To speak of a possible case means to speak of a kind of objects which is possibly non-empty.

All *logically possible* cases come into consideration for the determination of intensions. This includes also those cases that are causally impossible, i.e., excluded by the laws of nature holding in our universe, and certainly those that are excluded by laws which Karl believes to hold. Thus, if Karl believes that all P are Q by a law of nature, the linguist will still induce him to consider things that are P but not Q, and ask him whether or not he would apply to them the predicate under investigation (e.g., 'Pferd').

The inadequacy of the extensionalist thesis is also shown by the following example. Consider, on the one hand, these customary entries in German-English dictionaries:

(3) *Einhorn*, unicorn. *Kobold*, goblin,

and, on the other hand, the following unusual entries:

(4) *Einhorn*, goblin. *Kobold*, unicorn.

Now the two German words (and likewise the two English words) have the same extension, viz., the null class. Therefore, if the extensionalist thesis were correct, there would be no essential, empirically testable difference between (3) and (4). The extensionalist is compelled to say that the fact that (3) is generally accepted and (4) generally rejected is merely due to a tradition created by the lexicographers, and that there are no facts of German language behavior which could be regarded as evidence in favor of (3) as against (4). I wonder whether any linguist would be willing to accept (4). Or, to avoid the possibly misguiding influence of the lexicographers' tradition, let us put the question this way: would a man on the street, who has learned both languages by practical use without lessons or dictionaries, accept as correct a translation made according to (4)?

In general terms, the determination of the intension of a predicate may start from some instances denoted by the predicate. The essential task is then to find out what variations of a given specimen in various respects (e.g., size, shape, color) are admitted within the range of the predicate. The intension of a predicate may be defined as its range, which comprehends those possible kinds of objects for which the predicate holds. In this investigation of intension, the linguist finds a new kind of vagueness, which may be called *intensional vagueness*. As mentioned above, the extensional vagueness of the word 'Mensch' is very small, at least in the accessible region. First, the intermediate zone among animals now living on earth is practically empty. Second, if the ancestors of man are considered, it is probably found that Karl cannot easily draw a line; thus there is an intermediate zone, but it is relatively small. However, when the linguist proceeds to the determination of the *intension* of the word 'Mensch', the situation is quite different. He has to test Karl's responses to descriptions of strange kinds of animals, say intermediate between man and dog, man and lion, man and hawk, etc. It may be that the linguist and Karl know that these kinds of animals have never lived on earth; they do not know whether or not these kinds will ever occur on earth or on any other planet in any galaxy. At any rate, this knowledge or ignorance is irrelevant for the determination of intension. But Karl's ignorance has the psychological effect that he has seldom if ever thought of these kinds (unless he happens to be a student of mythology or a science-fiction fan) and therefore never felt an urge to make up his mind to which of them to apply the predicate 'Mensch'. Consequently, the linguist finds in Karl's responses a large intermediate zone for this predicate, in other words, a high intensional vagueness. The fact that Karl has not made such decisions means that the intension of the word 'Mensch' for him is not quite clear even to himself, that he does not completely understand his own word. This lack of clarity does not bother him much because it holds only for aspects which have very little practical importance for him.

The extensionalist will perhaps reject as impracticable the described procedure for determining intensions because, he might say, the man on the street is unwilling to say anything about nonexistent objects. If Karl happens to be over-realistic in this way, the linguist could still resort to a lie, reporting, say, his alleged observations of unicorns. But this is by no means necessary. The tests concerning intensions are independent of questions of existence. The man on the street is very well able to understand and to answer questions about assumed situations, where it is left open whether anything of the kind described will ever actually occur or not, and even about nonexisting situations. This is shown in ordinary conversations about alternative plans of action, about the truth of reports, about dreams, legends, and fairy tales.

Although I have given here only a rough indication of the empirical procedure for determining intensions, I believe that it is sufficient to make clear that it would be possible to write along the lines indicated a manual for determining intensions or, more exactly, for testing hypotheses concerning intensions. The kinds of rules in such a manual would not be essentially different from those customarily given for procedures in psychology, linguistics, and anthropology. Therefore the rules could be understood and carried out by any scientist (provided he is not infected by philosophical prejudices).[4]

[4] After writing the present paper I have become acquainted with a very interesting new book by Arne Naess [Interpretation]. This book describes in detail various procedures for test-

4. Intensions in the Language of Science

The discussions in this paper concern in general a simple, pre-scientific language, and the predicates considered designate observable properties of material bodies. Let us now briefly take a look at the *language of science.* It is today still mainly a natural language (except for its mathematical part), with only a few explicitly made conventions for some special words or symbols. It is a variant of the pre-scientific language, caused by special professional needs. The degree of precision is here in general considerably higher (i.e., the degree of vagueness is lower) than in the everyday language, and this degree is continually increasing. It is important to note that this increase holds not only for extensional but also for intensional precision; that is to say that not only the extensional intermediate zones (i.e., those of actual occurrences) but also the intensional ones (i.e., those of possible occurrences) are shrinking. In consequence of this development, also the intension concepts become applicable with increasing clarity. In the oldest books on chemistry, for example, there were a great number of statements describing the properties of a given substance, say water or sulphuric acid, including its reactions with other substances. There was no clear indication as to which of these numerous properties were to be taken as essential or definitory for the substance. Therefore, at least on the basis of the book alone, we cannot determine which of the statements made in the book were analytic and which synthetic for its author. The situation was similar with books on zoology, even at a much later time; we find a lot of statements, e.g., on the lion, without a clear separation of the definitory properties. But in chemistry there was an early development from the state described to states of greater and greater intensional precision. On the basis of the theory of chemical elements, slowly with increasing explicitness certain properties were selected as essential. For a compound, the molecular formula (e.g., 'H_2O') was taken as definitory, and later the molecular structure diagram. For the elementary substances, first certain experimental properties were more and more clearly selected

---

ing hypotheses concerning the synonymity of expressions with the help of questionnaires, and gives examples of statistical results found with these questionnaires. The practical difficulties and sources of possible errors are carefully investigated. The procedures concern the responses of the test persons, not to observed objects as in the present paper, but to pairs of sentences within specified contexts. Therefore the questions are formulated in the metalanguage, e.g., "Do the two given sentences in the given context express the same assertion to you?" Although there may be different opinions concerning some features of the various procedures, it seems to me that the book marks an important progress in the methodology of empirical meaning analysis for natural languages. Some of the questions used refer also to possible kinds of cases, e.g., "Can you imagine circumstances (conditions, situations) in which you would accept the one sentence and reject the other, or vice versa?" (p. 368). The book, both in its methodological discussions and in its reports on experiences with the questionnaires, seems to me to provide abundant evidence in support of the intensionalist thesis (in the sense explained in § 3 above).

as definitory, for example the atomic weight, later the position in Mendeleev's system. Still later, with a differentiation of the various isotopes, the nuclear composition was regarded as definitory, say characterized by the number of protons (atomic number) and the number of neutrons.

We can at the present time observe the advantages already obtained by the explicit conventions which have been made, though only to a very limited extent, in the language of empirical science, and the very great advantages effected by the moderate measure of formalization in the language of mathematics. Let us suppose—as I indeed believe, but that is outside of our present discussion—that this trend toward explicit rules will continue. Then the practical question arises whether rules of extension are sufficient or whether it would be advisable to lay down also rules of intension. In my view, it follows from the previous discussion that rules of intension are required, because otherwise intensional vagueness would remain, and this would prevent clear mutual understanding and effective communication.

5. The General Concept of the Intension of a Predicate

We have seen that there is an empirical procedure for testing, by observations of linguistic behavior, a hypothesis concerning the intension of a predicate, say 'Pferd', for a speaker, say Karl. Since a procedure of this kind is applicable to any hypothesis of intension, the general concept of the intension of any predicate in any language for any person at any time has a clear, empirically testable sense. This general concept of intension may be characterized roughly as follows, leaving subtleties aside: the intension of a predicate '*Q*' for a speaker *X* is the general condition which an object *y* must fulfil in order for *X* to be willing to ascribe the predicate '*Q*' to *y*. (We omit, for simplicity, the reference to a time *t*.) Let us try to make this general characterization more explicit. That *X* is able to use a language *L* means that *X* has a certain system of interconnected dispositions for certain linguistic responses. That a predicate '*Q*' in a language *L* has the property *F* as its intension for *X*, means that among the dispositions of *X* constituting the language *L* there is the disposition of ascribing the predicate '*Q*' to any object *y* if and only if *y* has the property *F*. (*F* is here always assumed to be an observable property, i.e., either directly observable or explicitly definable in terms of directly observable properties.) (The given formulation is oversimplified, neglecting vagueness. In order to take vagueness into account, a pair of intensions F_1, F_2 must be stated: *X* has the disposition of ascribing affirmatively the predicate '*Q*' to an object *y* if and only if *y* has F_1; and the disposition of denying '*Q*' for *y* if and only if y has F_2. Thus, if y has neither F_1 nor F_2, X will give neither an affirmative nor a negative response; the property of having neither F_1 nor F_2 constitutes the zone of vagueness, which may possibly be empty.)

The concept of intension has here been characterized only for thing-predicates. The characterization for expressions of other types, including sentences, can be given in an analogous way. The other concepts of the theory of intension can then be defined in the usual way; we shall state only those for 'synonymous' and 'analytic' in a simple form without claim to exactness.

Two expressions are *synonymous* in the language L for X at time t if they have the same intension in L for X at t.

A sentence is *analytic* in L for X at t if its intension (or range or truth-condition) in L for X at t comprehends all possible cases.

A language L was characterized above as a system of certain dispositions for the use of expressions. I shall now make some remarks on the *methodology of dispositional concepts*. This will help to a clearer understanding of the nature of linguistic concepts in general and of the concept of intension in particular. Let D be the disposition of X to react to a condition C by the characteristic response R. There are in principle, although not always in practice, two ways for ascertaining whether a given thing or person X has the disposition D (at a given time t). The first method may be called *behavioristic* (in a very wide sense); it consists in producing the condition C and then determining whether or not the response R occurs. The second way may be called the *method of structure analysis*. It consists in investigating the state of X (at t) in sufficient detail such that it is possible to derive from the obtained description of the state with the help of relevant general laws (say of physics, physiology, etc.) the responses which X would make to any specified circumstances in the environment. Then it will be possible to predict, in particular, whether, under the condition C, X would make the response R or not; if so, X has the disposition D, otherwise not. For example, let X be an automobile and D be the ability for a specified acceleration on a horizontal road at a speed of 10 miles per hour. The hypothesis that the automobile has this ability D may be tested by either of the following two procedures. The behavioristic method consists in driving the car and observing its performance under the specified conditions. The second method consists in studying the internal structure of the car, especially the motor, and calculating with the help of physical laws the acceleration which would result under the specified conditions. With respect to a psychological disposition and, in particular, a linguistic disposition of a person X, there is first the familiar behavioristic method and second, at least theoretically, the method of a micro-physiological investigation of the body of X, especially the central nervous system. At the present state of physiological knowledge of the human organism and especially the central nervous system, the second method is, of course, not practicable.

6. The Concept of Intension for a Robot

In order to make the method of structure analysis applicable, let us now consider the pragmatical investigation of the language of a robot rather than that of a human being. In this case we may assume that we possess much more detailed knowledge of the internal structure. The logical nature of the pragmatical concepts remains just the same. Suppose that we have a sufficiently detailed blueprint according to which the robot X was constructed and that X has abilities of observation and of use of language. Let us assume that X has three input organs A, B, and C, and an output organ. A and B are used alternatively, never simultaneously. A is an organ of visual observation of objects presented. B can receive a general description of a kind of object (a predicate expression) in the language L of X, which may consist of written marks or of holes punched in a card. C receives a predicate. These inputs constitute the question whether the object presented at A or any object satisfying the description presented at B is denoted in L for X by the predicate presented at C. The output organ may then supply one of three responses of X, for affirmation, denial, or abstention; the latter response would be given, e.g., if the observation of the object at A or the description at B is not sufficient to determine a definite answer. Just as the linguist investigating Karl begins with pointing to objects, but later, after having determined the interpretation of some words, asks questions formulated by these words, the investigator of X's language L begins with presenting objects at A, but later, on the basis of tentative results concerning the intensions of some signs of L, proceeds to present predicate expressions at B which use only those interpreted signs and not the predicate presented at C.

Instead of using this behavioristic method, the investigator may here use the method of structure analysis. On the basis of the given blueprint of X, he may be able to calculate the responses which X would make to various possible inputs. In particular, he may be able to derive from the given blueprint, with the help of those laws of physics which determine the functioning of the organs of X, the following result with respect to a given predicate 'Q' of the language L of X and specified properties F_1 and F_2 (observable for X): If the predicate 'Q' is presented at C, then X gives an affirmative response if and only if an object having the property F_1 is presented at A and a negative response if and only if an object with F_2 is presented at A. This result indicates that the boundary of the intension of 'Q' is somewhere between the boundary of F_1 and that of F_2. For some predicates the zone of indeterminateness between F_1 and F_2 may be fairly small and hence this preliminary determination of the intension fairly precise. This might be the case, for example, for color predicates if the investigator has a sufficient number of color specimens.

After this preliminary determination of the intensions of some predicates constituting a restricted vocabulary V by calculations concerning input A, the investigator will proceed to make calculations concerning descriptions containing the predicates of V to be presented at B. He may be able to derive from the blueprint the following result: If the predicate 'P' is presented at C, and any description D in terms of the vocabulary V is presented at B, X gives an affirmative response if and only if D (as interpreted by the preliminary results) logically implies G_1, and a negative response if and only if D logically implies G_2. This result indicates that the boundary of the intension of 'P' is between the boundary of G_1 and that of G_2. In this way more precise determinations for a more comprehensive part of L and finally for the whole of L may be obtained. (Here again we assume that the predicates of L designate observable properties of things.)

It is clear that the method of structure analysis, if applicable, is more powerful than the behavioristic method, because it can supply a general answer and, under favorable circumstances, even a complete answer to the question of the intension of a given predicate.

Note that the procedure described for input A can include empty kinds of objects and the procedure for input B even causally impossible kinds. Thus, for example, though we cannot present a unicorn at A, we can nevertheless calculate which response X would make if a unicorn were presented at A. This calculation is obviously in no way affected by any zoological fact concerning the existence or nonexistence of unicorns. The situation is different for a kind of objects excluded by a law of physics, especially, a law needed in the calculations about the robot. Take the law l_1: "Any iron body at 60° F is solid". The investigator needs this law in his calculation of the functioning of X, in order to ascertain that some iron cogwheels do not melt. If now he were to take as a premise for his derivation the statement "A liquid iron body having the temperature of 60° F is presented at A", then, since the law l_1 belongs also to his premises, he would obtain a contradiction; hence every statement concerning *X*'s response would be derivable, and thus the method would break down. But even for this case the method still works with respect to *B*. He may take as premise "The description 'liquid iron body with the temperature of 60° F' (that is, the translation of this into *L*) is presented at *B*". Then no contradiction arises either in the derivation made by the investigator or in that made by *X*. *The derivation carried out by the investigator* contains the premise just mentioned, which does not refer to an iron body but to a description, say a card punched in a certain way; thus there is no contradiction, although the law l_1 occurs also as a premise. On the other hand, in *the derivation made by the robot X*, the card presented at *B* supplies, as it were, a premise of the form "*y* is a liquid iron body at 60° F"; but here the law l_1 does not occur as a premise, and thus no contradiction occurs. *X* makes merely logical deductions from the one premise stated and, if the predicate '*R*' is presented at *C*, tries to come either to the conclusion "*y* is *R*" or "*y* is not *R*". Suppose the investigator's calculation leads to the result that *X* would derive the conclusion "*y* is *R*" and hence that *X* would give an affirmative response. This result would show that the (causally impossible) kind of liquid iron bodies at 60° F is included in the range of the intension of '*R*' for *X*.

I have tried to show in this paper that in a pragmatical investigation of a natural language there is not only, as generally agreed, an empirical method for ascertaining which objects are denoted by a given predicate and thus for determining the extension of the predicate, but also a method for testing a hypothesis concerning its intension (designative meaning).[5] The intension of a predicate for a speaker *X* is, roughly speaking, the general condition which an object must fulfil for *X* to be willing to apply the predicate to it. For the determination of intension, not only actually given cases must be taken into consideration, but also possible cases, i.e., kinds of objects which can be described without self-contradiction, irrespective of the question whether there are any objects of the kinds described. The intension of a predicate can be determined for a robot just as well as for a

[5] Y. Bar-Hillel in a recent paper [Syntax] defends the concept of meaning against those contemporary linguists who wish to ban it from linguistics. He explains this tendency by the fact that in the first quarter of this century the concept of meaning was indeed in a bad methodological state; the usual explanations of the concept involved psychologistic connotations, which were correctly criticized by Bloomfield and others. Bar-Hillel points out that the semantical theory of meaning developed recently by logicians is free of these drawbacks. He appeals to the linguists to construct in an analogous way the theory of meaning needed in their empirical investigations. The present paper indicates the possibility of such a construction. The fact that the concept of intension can be applied even to a robot shows that it does not have the psychologistic character of the traditional concept of meaning.

human speaker, and even more completely if the internal structure of the robot is sufficiently known to predict how it will function under various conditions. On the basis of the concept of intension, other pragmatical concepts with respect to natural languages can be defined, synonymy, analyticity, and the like. The existence of scientifically sound pragmatical concepts of this kind provides a practical motivation and justification for the introduction of corresponding concepts in pure semantics with respect to constructed language systems.

E. ON SOME CONCEPTS OF PRAGMATICS

In an earlier paper [Synonymy], I discussed the pragmatical concept of intension in order to defend its scientific legitimacy. I gave only an informal analysis, not an exact explication. Chisholm [Note] is certainly right in saying that my account was an oversimplification. But this was intentional; in particular, I deliberately left aside not only the possible effects of vagueness, but also those of factual errors of the speaker Karl (see my references to these errors as possibly falsifying the linguist's results, at the end of § 2). I would agree with Chisholm in preferring the third of the three ways for refining the analysis, the one using the concept of belief.

It seems that a more thorough analysis of intension, belief, and related concepts would require a conceptual framework of theoretical pragmatics. I shall mention here a few concepts that might come into consideration as a basis of such a framework. I will state merely the general form and roughly indicate the meaning of these concepts without attempting any analysis.

I think today that the basic concepts of pragmatics are best taken, not as behavioristically defined disposition concepts of the observation language, but as theoretical constructs in the theoretical language, introduced on the basis of postulates and connected with the observation language by rules of correspondence. The concept of *belief* is sometimes construed, e.g., by Church, as a relation between a person and a proposition. I previously made an attempt at explicating it as a relation between a person and a sentence. Perhaps both concepts are useful; the first is non-pragmatical; it characterizes a state of a person not necessarily involving language. The second concept is pragmatical. Let us write 'B' for the first, 'T' for the second. Let a sentence of the form

$$B(X,t,p) \tag{1}$$

say that the person X at the time t believes that p. This is understood in a weak sense, as not implying either that X is aware of the belief or that he is able to verbalize it. Let a sentence of the form

$$T(X,t,S,L) \tag{2}$$

say that X at t takes the sentence S of the language L to be true (consciously or not). For the sake of simplicity, I take here both B and T as simple relations. In a more adequate systematization, both should be taken as concepts of degree. Now the pragmatical concept of *intension* serves as a connecting link between B and T. Let a sentence of the form

(3) $$Int(p,S,L,X,t)$$

say that the proposition p is the intension of the sentence S in the language L for X at t. (Another alternative would take "sense", as used by Church, instead of "intension". In either case, the sentences (1) and (3) are non-extensional. I do not think that there is any compelling reason for avoiding the use of an intensional language for science, because such a language can be completely translated into an extensional one, as I shall show elsewhere.) If suitable postulates and rules for the three concepts are laid down, (2) can presumably be inferred from (1) and (3) (either deductively or inductively), and (1) from (2) and (3).

Since T refers only to sentences, pragmatics needs a concept of intension primarily for sentences. But the concept of intension for other types of designators is essential too. In any language, the intension of a compound sentence is a function of the intensions of its parts. It is only due to this fact that a user of a language is able to understand an unlimited number of sentences on the basis of his understanding of a limited number of words or phrases.

Pragmatics needs in addition one or two concepts of *utterance*. Let

(4) $$A(X,t,S,L)$$

mean that X at t wills deliberately to utter a token of S as a sentence of the language L in the sense of an *assertion*. Since the concept A involves the purpose or intention, it is clearly a theoretical construct. The following concept, on the other hand, belongs to the observation language. Let

(5) $$U(X,t,R)$$

mean that X at t produces with his speaking organs a series of audible sounds R. Suppose that R is a token of S:

(6) $$U(X,t,S).$$

This sentence contains no reference to L. The fact that the sounds S are meant by X as a sentence of L, is not directly observed, but can at best be inferred inductively. The rules of correspondence may supply a connection between A and U. Suppose that (6) is established as a result of observations. Then it may be possible to infer inductively, with the help of suitable auxiliary premises concerning the "normality" of the situation and previously confirmed facts about X including (3), first (4), then (2), and finally (1).

There is an urgent need for a system of theoretical pragmatics, not only for psychology and linguistics, but also for analytic philosophy. Since pure semantics is sufficiently developed, the time seems ripe for attempts at constructing tentative outlines of pragmatical systems. Such an outline may first be restricted to small groups of concepts (e.g., those of belief, assertion, and utterance); it may then be developed to include all those concepts needed for discussions in the theory of knowledge and the methodology of science.

BIBLIOGRAPHY

[The abbreviated titles in brackets are used in citations throughout this book.]

Bar-Hillel, Yehoshua. [Syntax]. Logical syntax and semantics, *Language*, 30 (1954), 230–37.

Bennett, Albert A., and **Baylis, Charles A.** [Logic]. *Formal logic: A modern introduction*. New York, 1939.

Bergmann, Gustav. Two cornerstones of empiricism, *Synthese*, 8 (1953), 435–52.

Carnap, Rudolf. [Syntax]. *Logical syntax of language*. Orig., Vienna, 1934; English translation, London and New York, 1937.

———. [I]. *Introduction to semantics*. Studies in Semantics, Vol. I. Cambridge, Mass., 1942.

———. [II]. *Formalization of logic*. Studies in Semantics, Vol. II. Cambridge, Mass., 1943.

———. [Inductive]. On inductive logic, *Phil. Science*, 12 (1945), 72–97.

———. [Remarks]. Remarks on induction and truth, *Phil. and Phenom. Res.*, 6 (1946), 590–602. (§ 3 of this paper was reprinted in [Truth].)

———. [Modalities]. Modalities and quantification, *J. Symb. Logic*, 11 (1946), 33–64.

———. [Truth]. Truth and confirmation. In: Feigl and Sellars [Readings].

———. A reply to Leonard Linsky, *Phil. Science*, 16 (1949), 347–50.

*———. Empiricism, semantics, and ontology, *Revue Intern. de Phil.*, 4 (1950), 20–40.

———. [Probability]. *Logical foundations of probability*. Chicago, 1950. (§§ 2–6 on explication, §§ 17–20 on truth and L-concepts.)

*———. [Postulates]. Meaning postulates, *Phil. Studies*, 3 (1952), 65–73.

———. *Einführung in die symbolische Logik, mit besonderer Berücksichtigung ihrer Anwendungen*. Wien, 1954. (English translation in preparation.)

*———. On belief sentences. Reply to Alonzo Church. In: Macdonald [Philosophy], pp. 128–31.

*———. [Synonymy]. Meaning and synonymy in natural languages, *Phil. Studies*, 7 (1955), 33–47.

*———. On some concepts of pragmatics, *Phil. Studies*, 6 (1955), 89–91.

Chisholm, R. M. [Note]. A note on Carnap's meaning analysis, *Phil. Studies*, 6 (1955), 87–89.

Church, Alonzo. [Dictionary]. Articles in D. D. Runes (ed.), *The Dictionary of Philosophy*. New York, 1942.

———. [Review C.]. Carnap's introduction to semantics (a review of Carnap [I]), *Phil. Review*, 52 (1943), 298–304.

———. [Review Q.]. A review of Quine [Notes], *J. Symb. Logic*, 8 (1943), 45–47.

———. [Belief]. On Carnap's analysis of statements of assertion and belief, *Analysis*, 10 (1950), 97–99. Reprinted in: Macdonald [Philosophy].

———. A formulation of the logic of sense and denotation. In: P. Henle (ed.), *Essays in honor of Henry Sheffer*, pp. 3–24. New York, 1951.

———. The need for abstract entities in semantic analysis, *Proc. Amer. Acad. of Arts and Sciences*, 80 (1951), 100–112.

———. Intentional isomorphism and identity of belief, *Phil. Studies*, 5 (1954), 65–73.

Feigl, Herbert, and **Sellars, Wilfrid.** [Readings]. *Readings in philosophical analysis*. New York, 1949.

* The starred articles are *reprinted in the Supplement to this book*.

Frege, Gottlob. [Grundlagen]. *Die Grundlagen der Arithmetik.* (1884) 1934. Reprinted with English translation in: J. L. AUSTIN (ed.), *Foundations of arithmetic.* New York, 1950.

———. [Sinn]. Ueber Sinn und Bedeutung, *Zeitschr. für Philos. und philos. Kritik*, 100 (new ser., 1892), 25–50. (English translations in: FEIGL and SELLARS [Readings] and in [Translations].)

———. [Grundgesetze]. *Grundgesetze der Arithmetik*, Vols. I, II. Jena, 1893; 1903.

———. [Translations]. *Translations from the philosophical writings.* Translated by P. GEACH and M. BLACK. Oxford, 1952.

Gewirth, Alan. The distinction between analytic and synthetic truths, *J. Phil.*, 50 (1953), 397–425.

Hilbert, David, and **Bernays, Paul.** [Grundlagen I]. *Grundlagen der Mathematik*, Vol. I. Berlin, 1934.

Kemeny, John. A new approach to semantics. Part I, *J. Symb. Logic*, Vol. 21 (1956). (Further parts in preparation.)

Lewis, C. I. *A survey of symbolic logic.* Berkeley, 1918.

———. [Meaning]. The modes of meaning, *Phil. and Phenom. Res.*, 4 (1943–44), 236–50. Reprinted in: LINSKY [Semantics].

———. *An analysis of knowledge and valuation.* La Salle, Ill., 1946.

Lewis, C. I., and **Langford, C. H.** *Symbolic logic.* New York, 1946.

Linsky, Leonard. Some notes on Carnap's concept of intensional isomorphism and the paradox of analysis, *Phil. Science*, 16 (1949), 347–50.

——— (ed.). [Semantics]. *Semantics and the philosophy of language.* Urbana, 1952.

———. Description and the antinomy of the name-relation, *Mind*, 61 (1952), 273–75.

Macdonald, Margaret (ed.). [Philosophy]. *Philosophy and analysis: A selection of articles published in Analysis.* Oxford, 1954.

Martin, Richard M. [Analytic]. On 'analytic', *Phil. Studies*, 3 (1952), 42–47.

Mates, Benson. [Analytic]. Analytic sentences, *Phil. Review*, 60 (1951), 525–34.

———. [Synonymity]. Synonymity. In: *Meaning and interpretation.* Univ. of Cal. Publications in Philos., 25 (1950), 201–26. Reprinted in: LINSKY [Semantics].

Morris, Charles W. [Signs]. *Signs, language, and behavior.* New York, 1946.

Naess, Arne. [Interpretation]. *Interpretation and preciseness: A contribution to the theory of communication.* Skrifter Norske Vid. Akademi, Oslo, II. Hist.-Filos. Klasse, 1953, No. 1.

Nagel, Ernest. Logic without ontology. In: KRIKORIAN (ed.), *Naturalism and the human spirit.* New York, 1944. Reprinted in: FEIGL and SELLARS [Readings].

———. [Review C.]. (Review of the first edition of this book), *J. Phil.*, 45 (1948), 467–72.

Putnam, Hilary. [Synonymity]. Synonymity and the analysis of belief sentences, *Analysis*, 14 (1954), 114–22.

Quine, W. V. [Designation]. Designation and existence, *J. Phil.*, 36 (1939), 702–9.

———. [M.L.]. *Mathematical logic.* New York, 1940.

———. [Notes]. Notes on existence and necessity, *J. Phil.*, 40 (1943), 113–27. Reprinted in: LINSKY [Semantics].

———. [Universals]. On universals, *J. Symb. Logic*, 12 (1947), 74–84.

———. [What]. On what there is, *Review of Metaphysics*, 2 (1948), 21–38. Reprinted in [Logical] and in: LINSKY [Meaning].

———. On Carnap's views on ontology, *Phil. Studies*, 2 (1951), 65–72.

———. [Semantics]. Semantics and abstract objects, *Proc. Amer. Acad. of Arts and Sciences*, 80 (1951), 90–96. Partly reprinted in [Logical].

———. [Dogmas]. Two dogmas of empiricism, *Phil. Review*, 60 (1951), 20–43. Reprinted in [Logical].

———. [Logical]. *From a logical point of view: Nine logico-philosophical essays.* Cambridge, Mass., 1953.

Russell, Bertrand. [Denoting]. On denoting, *Mind*, 14 (1905), 479–93. Reprinted in: FEIGL and SELLARS [Readings].

———. [P.M.]. *See* WHITEHEAD.

———. [Inquiry]. *An inquiry into meaning and truth.* New York, 1940.

Ryle, Gilbert. [Meaning]. Meaning and necessity, *Philosophy*, 24 (1949), 69–76.

Scheffler, Israel. On synonymy and indirect discourse, *Phil. Science*, 22 (1955), 39–44.

Sellars, Wilfrid. Putnam on synonymity and belief, *Analysis*, 15 (1955), 117–20.

Smullyan, Arthur F. Modality and description, *J. Symb. Logic*, 13 (1948), 31–37.

Tarski, Alfred. [Wahrheitsbegriff]. Der Wahrheitsbegriff in den formalisierten Sprachen, *Studia philosophica*, I (1936), 261–405. Originally published in 1933.

———. [Truth]. The semantic conception of truth and the foundations of semantics, *Phil. and Phenom. Res.*, 4 (1944), 341–76. Reprinted in: LINSKY [Semantics] and in: FEIGL and SELLARS [Readings].

———. *Logic, semantics, and metamathematics.* Oxford, 1955.

White, Morton. [Analytic]. The analytic and the synthetic: An untenable dualism. In: SIDNEY HOOK (ed.), *John Dewey*. New York, 1950. Reprinted in: LINSKY [Semantics].

Whitehead, A. N., and **Russell, B.** [P.M.]. *Principia mathematica.* 3 vols. Cambridge, England, 1910–13; 2d ed., 1925–27.

Wilson, Neil L. Designation and description, *J. Philos.*, 50 (1953), 369–83.

Wittgenstein, Ludwig. [Tractatus]. *Tractatus logico-philosophicus.* London, 1922.

INDEX

[The numbers refer to pages. The most important terms, names, and references are indicated by boldface type.]

SYMBOLS

PHOENIX BOOKS
Literature and Language

P 15 *R. S. Crane,* EDITOR: Critics and Criticism (Abridged)
P 26 *David Cecil:* Victorian Novelists
P 37 *Charles Feidelson, Jr.:* Symbolism and American Literature
P 38 *R. W. B. Lewis:* The American Adam: Innocence, Tragedy, and Tradition in the Nineteenth Century
P 50, P 51 *Harold C. Goddard:* The Meaning of Shakespeare (2 vols.)
P 52 *Zellig S. Harris:* Structural Linguistics
P 54 *Grover Smith:* T. S. Eliot's Poetry and Plays
P 60 *E. H. Sturtevant:* Linguistic Change: An Introduction to the Historical Study of Language
P 68 *Rosemond Tuve:* Elizabethan and Metaphysical Imagery: Renaissance Poetic and 20th Century Critics
P 72 *Elder Olson:* The Poetry of Dylan Thomas
P 77 *Edward W. Rosenheim, Jr.:* What Happens in Literature
P 91 *Paul Goodman:* The Structure of Literature
P 93 *Wallace Fowlie:* Mallarmé
P 106 *John T. Waterman:* Perspectives in Linguistics
P 109 *I. J. Gelb:* A Study of Writing
P 119 *Joseph H. Greenberg:* Essays in Linguistics
P 123 *Mitford McLeod Mathews:* Beginnings of American English: Essays and Comments
P 127 *Milton Crane:* Shakespeare's Prose
P 132 *Merlin Bowen:* The Long Encounter: Self and Experience in the Writings of Herman Melville
P 140 *Richard D. Altick:* The English Common Reader: A Social History of the Mass Reading Public, 1800–1900
P 148 *Ernest Hatch Wilkins:* Life of Petrarch
P 158 *Victor Brombert:* The Intellectual Hero: Studies in the French Novel, 1880–1955
P 163 *Ernest K. Bramsted:* Aristocracy and the Middle-Classes in Germany: Social Types in German Literature, 1830–1900
P 169 *Arthur Ryder,* TRANSLATOR: The Panchatantra
P 178 *Henry M. Hoenigswald:* Language Change and Linguistic Reconstruction
P 180 *David Daiches:* The Novel and the Modern World
P 188 *Chester E. Eisinger:* Fiction of the Forties
P 191 *Dennis Ward:* The Russian Language Today

P 200 *Samuel Daniel:* Poems and A Defence of Ryme

P 212 *James J. Y. Liu:* The Art of Chinese Poetry

P 213 *Murray Krieger:* The Tragic Vision

P 217 *Knox C. Hill:* Interpreting Literature

P 224 *James E. Miller, Jr.:* A Critical Guide to *Leaves of Grass*

P 229 *Mitford Mathews:* Americanisms

P 230 *André Martinet:* Elements of General Linguistics

P 234 *George Williamson:* The Senecan Amble: A Study in Prose Form from Bacon to Collier

P 239 *Robert B. Heywood,* EDITOR: The Works of the Mind

P 250 *Sherwood Anderson:* Windy McPherson's Son

P 251 *Henry B. Fuller:* With the Procession

P 252 *Ring Lardner:* Gullible's Travels, Etc.

P 267 *Wayne C. Booth:* The Rhetoric of Fiction

P 268 *Jean H. Hagstrum:* Samuel Johnson's Literary Criticism

PHOENIX BOOKS
in Philosophy

P 1 *Ernst Cassirer, Paul Oskar Kristeller, and John Herman Randall, Jr.:* The Renaissance Philosophy of Man
P 5 *Jacques Maritain:* Man and the State
P 8 *T. V. Smith,* EDITOR: Philosophers Speak for Themselves: From Thales to Plato
P 9 *T. V. Smith,* EDITOR: Philosophers Speak for Themselves: From Aristotle to Plotinus
P 16 *Karl Löwith:* Meaning in History: The Theological Implications of the Philosophy of History
P 17 *T. V. Smith and Marjorie Grene,* EDITORS: Philosophers Speak for Themselves: From Descartes to Locke
P 18 *T. V. Smith and Marjorie Grene,* EDITORS: Philosophers Speak for Themselves: Berkeley, Hume, and Kant
P 30 *Rudolph Carnap:* Meaning and Necessity: A Study in Semantics and Modal Logic
P 34 *Marjorie Grene:* Introduction to Existentialism
P 44 *Richard Weaver:* Ideas Have Consequences
P 67 *Yves R. Simon:* Philosophy of Democratic Government
P 81 *Hans Reichenbach:* Experience and Prediction: An Analysis of the Foundations and Structure of Knowledge
P 110 *Manley Thompson:* The Pragmatic Philosophy of C. S. Peirce
P 112 *Leo Strauss:* The Political Philosophy of Hobbes: Its Basis and Its Genesis
P 114 *Lewis White Beck:* A Commentary on Kant's Critique of Practical Reason
P 128 *Michael Polanyi:* The Study of Man
P 142 *Charles Hartshorne and William L. Reese,* EDITORS: Philosophers Speak of God
P 143 *Karl Jaspers:* The Future of Mankind
P 155 *Michael Polanyi:* Science, Faith and Society
P 167 *Muhsin Mahdi:* Ibn Khaldûn's Philosophy of History
P 184 *Paul Shorey:* What Plato Said
P 195 *Leo Strauss:* Natural Right and History
P 198 *Émile Bréhier:* The Hellenic Age
P 199 *Émile Bréhier:* The Hellenistic and Roman Age
P 201 *David Grene:* Greek Political Thought
P 225 *N. A. Nikam and Richard McKeon,* EDITORS: The Edicts of Asoka
P 239 *Robert B. Heywood,* EDITOR: The Works of the Mind
P 264 *Émile Bréhier:* The Middle Ages and the Renaissance
P 265 *A. C. Ewing:* A Short Commentary on Kant's Critique of Pure Reason
P 266 *Marjorie Grene:* A Portrait of Aristotle
P 269 *Arthur F. Wright,* EDITOR: Studies in Chinese Thought
P 270 *John K. Fairbank,* EDITOR: Chinese Thought and Institutions
P 272 *George H. Mead:* Mind, Self, and Society